MEASUREMENT TECHNIQUES
FOR POLYMERIC SOLIDS

This volume is based on papers presented at a conference on Measurement Techniques for Polymeric Solids held in December 1982 at the National Physical Laboratory, Teddington, Middlesex, UK.

MEASUREMENT TECHNIQUES FOR POLYMERIC SOLIDS

Edited by

R. P. BROWN

*Rubber and Plastics Research Association, Shawbury,
Shrewsbury, UK*

and

B. E. READ

National Physical Laboratory, Teddington, Middlesex, UK

Reprinted from the Journal
Polymer Testing
Vol. 4, Nos 2–4, 1984

ELSEVIER APPLIED SCIENCE PUBLISHERS

LONDON and NEW YORK

ELSEVIER APPLIED SCIENCE PUBLISHERS LTD
Ripple Road, Barking, Essex, England

Sole Distributor in the USA and Canada
ELSEVIER SCIENCE PUBLISHING CO., INC.
52 Vanderbilt Avenue, New York, NY 10017, USA

British Library Cataloguing in Publication Data

Measurement techniques for polymeric solids.
   1. Polymers and polymerization—Measurement
   I. Brown, R. P. II. Read, B. E.
   620.1′92′0287   TA455.P58

   ISBN 085334-274-1

WITH 10 TABLES AND 80 ILLUSTRATIONS

© ELSEVIER APPLIED SCIENCE PUBLISHERS LTD 1984

© CROWN COPYRIGHT 1984—pp. 101–115, 143–164,
193–194, 225–249, 253–272

Printed in Northern Ireland by The Universities Press, Belfast

Preface

The National Physical Laboratory (NPL) Conference on *Measurement Techniques for Polymeric Solids* was arranged in December 1982 to review techniques for quantitatively characterising the structures and physical behaviour of plastics and rubbers, and to discuss the relevance of the methods to industrial needs. The first day of this two-day meeting was devoted to a discussion of the requirements for physical testing and of techniques for determining polymer structure, internal stress and dimensional changes. During the second day, emphasis was given to the measurement of relevant mechanical and electrical properties.

In this volume 12 of the 16 invited papers are published in full and summaries are included of the other presentations. The wide range of techniques discussed yield data relevant to the selection of materials, the design of components and the prediction of their service performance. A few of the methods are now well established and form the basis of international Standards, but many have recently evolved from polymer research investigations and have potential for providing future improvements to existing Standards.

An exhibition of relevant measuring equipment formed an important part of the meeting and I would like to take this opportunity to thank the many participating organisations for their contributions. Encouragement and support for the meeting was also given by the Plastics and Rubber Institute, the Institute of Physics, the Macro Group UK and the Polymer Engineering Directorate. The collaboration of the Editor of *Polymer Testing* in publishing these proceedings is gratefully acknowledged.

B. E. Read
National Physical Laboratory,
Teddington, Middlesex TW11 0LW, UK

Contents

List of Contributors

M. J. BEVIS
   *Department of Non-Metallic Materials, Brunel University, Ux-
   bridge, Middlesex, UK*

A. R. BLYTHE
   *BICC Research and Engineering Ltd, Wood Lane, London W12
   7DX, UK*

J. BOWMAN
   *Department of Non-Metallic Materials, Brunel University, Ux-
   bridge, Middlesex, UK*

R. P. BROWN
   *Rubber and Plastics Research Association of Great Britain,
   Shawbury, Shrewsbury, Shropshire SY4 4NR, UK*

G. D. DEAN
   *National Physical Laboratory, Teddington, Middlesex TW11
   0LW, UK*

J. C. DUNCAN
   *National Physical Laboratory, Teddington, Middlesex TW11
   0LW, UK*

A. E. ENNOS
   *National Physical Laboratory, Teddington, Middlesex TW11
   0LW, UK*

J. ESS
   *Department of Non-Metallic Materials, Brunel University, Ux-
   bridge, Middlesex, UK*

P. S. GILL
   *Du Pont (UK) Ltd, Wedgwood Way, Stevenage, Hertfordshire
   SG1 4QN, UK*

M. J. GREGORY

The Malaysian Rubber Producers' Research Association, Tun Abdul Razak Laboratory, Brickendonbury, Hertfordshire SG13 8NL, UK

P. R. HORNSBY

Department of Non-Metallic Materials, Brunel University, Uxbridge, Middlesex, UK

A. F. JOHNSON

National Physical Laboratory, Teddington, Middlesex TW11 0LW, UK

C. C. LAWRENCE

North East London Polytechnic, Manufacturing Studies, Longbridge Road, Dagenham, Essex, UK

J. D. LEAR

Du Pont (UK) Ltd, Wedgwood Way, Stevenage, Hertfordshire SG1 4QN, UK

J. N. LECKENBY

Du Pont (UK) Ltd, Wedgwood Way, Stevenage, Hertfordshire SG1 4QN, UK

S. LIN

Department of Non-Metallic Materials, Brunel University, Uxbridge, Middlesex, UK

L. N. McCARTNEY

National Physical Laboratory, Teddington, Middlesex TW11 0LW, UK

D. E. MEYER

National Physical Laboratory, Teddington, Middlesex TW11 0LW, UK

D. R. MOORE

Imperial Chemical Industries PLC, Petrochemicals and Plastics

Division, Wilton Centre, PO Box 80, Wilton, Middlesbrough, Cleveland TS6 8JE, UK

B. E. READ
National Physical Laboratory, Teddington, Middlesex TW11 0LW, UK

M. J. RICHARDSON
National Physical Laboratory, Teddington, Middlesex TW11 0LW, UK

A. STEVENSON
The Malaysian Rubber Producers' Research Association, Tun Abdul Razak Laboratory, Brickendonbury, Hertfordshire SG13 8NL, UK

K. THOMAS
National Physical Laboratory, Teddington, Middlesex TW11 0LW, UK

D. VESELY
Department of Non-Metallic Materials, Brunel University, Uxbridge, Middlesex, UK

R. E. WETTON
Department of Chemistry, Loughborough University of Technology, Loughborough, Leicestershire LE11 3TU, UK

J. R. WHITE
Department of Metallurgy and Engineering Materials, University of Newcastle upon Tyne, Newcastle upon Tyne NE1 7RU, UK

Polymer Testing **4** (1984) 91–100

Requirements for Physical Testing of Rubbers and Plastics

R. P. Brown

Rubber and Plastics Research Association of Great Britain,
Shawbury, Shrewsbury, Shropshire SY4 4NR, UK

SUMMARY

This introductory paper reviews in a general way the requirements for physical testing of rubbers and plastics. The perennial questions of why and how we test are raised and considered with reference to product examples. A grouping of physical tests is suggested based on the use and value of the information they provide and related to the reasons for testing. Particular attention is given to the role of standards and to the sometimes conflicting requirements of quality control and design.

1. INTRODUCTION

Almost all the papers at this conference deal either with specific techniques or with particular groups of parameters. Clearly, no one paper, nor even one conference, can fully cover all the physical tests which for one reason or another are made on polymers. The objective of this introductory paper is to review in a general way the requirements for physical testing of rubber and plastics, and hence help to set the scene in which the following papers give detailed consideration to particular elements.

It is also the aim of this paper to pose questions and stimulate thought on the basic questions of the needs for testing and our approaches to satisfying them.

91

Polymer Testing 0142-9418/84/$03·00 © Elsevier Applied Science Publishers Ltd, England, 1984. Printed in Northern Ireland

We can state the old boring questions:

Why are we testing?
Which properties should we measure?
How should we test?
What are the problems in testing, how do we overcome them?

One selection of answers to 'Why are we testing?' might be:

Quality control
Design data
Predicting service performance
Investigating failures

Generally speaking, these reasons will apply to any material or product and I would further suggest that, because the requirements for each of these reasons is different, it is not sensible to answer the other questions of what properties and how we should test until the reason for testing has been clearly identified. This is of course an obvious point but I emphasise it because in my experience communication in the testing world between, for example, the university research worker and the factory floor quality controller is often limited by lack of appreciation of why the other person is testing and what he needs to get from the tests.

This leads to the question, in addition to 'Why are we testing?', 'What must we get out of the results?' This is an important question because it implies that whatever the reason for testing, the answers to which properties and which methods will to some extent depend on the level of information we need. It is easy to envisage that in two cases, both with the same reason for testing, there may be two different levels of need—for example, in quantity, depth or precision of data.

2. PRODUCT EXAMPLES

An illustration or example is useful to focus the mind and I would like to refer to a random selection of four products to briefly consider which properties we measure, what methods we use and what problems we have—considering just one reason for testing for each product.

2.1. The plastic bath (service performance)

What properties of a bath do people consider? Certainly the appearance, possibly the staining and scratch resistance, almost certainly not the stiffness of the material and its resistance to distortion under load and heat. Despite not consciously thinking of such performance parameters, it would be reasonable to say that over say 6 months of use the bath will have been adequately tested for service performance.

An important question is whether to test the product or to test the material from which it is made. If we consider the bath from the point of view of testing for fitness for purpose it will clearly be reasonable to test the product. It is not a prohibitively expensive object and it is easy to visualise the sort of proving routine that is needed.

It may be instructive to look at specifications for plastic baths. Although specifications (as opposed to codes of practice) are rarely, if ever, helpful from a design data generation point of view, they should contain tests which are relevant for the other reasons for testing listed. There is a draft CEN specification for acrylic baths and also a British Standard, and probably others. They are not identical although roughly speaking they both contain the sort of tests on the actual baths which one would instinctively think were relevant to performance—loading tests to check deflection, resistance to hot water and household chemicals, etc.

It is not intended to consider the properties or methods in detail here but to raise questions of principle—we have a bath and we are testing for fitness for purpose. On the plus side we have a product we can afford to test in the complete form and it should not be too difficult to list the properties of importance. What are the problems?

Even for such a common homely product there is not, it would seem, universal agreement as to the test methods to be used. I suggest we have a human problem of far too great a delay in getting together and defining what tests should be used rather than insurmountable technical barriers to designing adequate tests. Experience at RAPRA has also shown that certain of the tests do require refining to perform properly, but this again is mostly due to too little effort to date in proving the methods.

There is of course a second problem in that the bath will be expected to last for many years and the tests can only be short-term. Are the extrapolation rules valid? The answer to this well-known constantly occurring question is often no, although in the case of the bath the performance tests used appear to give confidence of a reasonable service life.

2.2. The electrician's glove (quality control)

An electrician at the top of a telegraph pole puts a lot of faith in the thin piece of rubber from which his glove is made. He will be very concerned that the quality control on the insulating glove is without fault.

In this case the properties of importance are clear; insulation level, no flaws, and adequate tear resistance or strength. The test methods and the quality assurance (QA) schedules are well established in standards, the most important is 100% testing for electrical breakdown strength and current leakage. This is a particularly good test because it is a proof test and non-destructive to a good glove.

It might be said that there are no problems in this case—we know the properties and the test methods have been defined. However, the tests are relatively slow and expensive, and ideally quality-control tests should be rapid as well as adequately screening bad production.

In theory at least, there is room for the development of instrumentation and methods to give more cost-effective testing. Probably the routine would have been speeded up for a less safety critical product but here very convincing proof would be needed that any quicker but less directly relevant service test was adequate. Nevertheless, it would be comforting to have a test that could be used in the field immediately before use.

2.3. The GRP motor boat (design data)

If you had bought such an expensive boat you would doubtless wish to be assured that it had been well designed. Even if we are not concerned with its sailing performance or non-polymer parts there is still a complex structure which costs a lot of money. It would not be reasonable to get it right by trial and error, hence it is essential to have reliable data on the materials employed which can be used for design purposes.

Again, it is not appropriate to detail here the properties and test methods which would be needed. Suffice to say that the product will be subjected to complex stress patterns at various loading rates whilst being subject to multiple environmental attack. This case will serve to point out the problems:

Rules are needed to relate strength and stiffness tests on laboratory samples to a complex shape.

Rules are needed to relate this data to both dynamic conditions and long times.

Rules are needed to superimpose the effects of environment onto the mechanical performance data.

These represent the needs for all design data and imply considerable understanding of material behaviour and the existence of test methods yielding data which can be extrapolated to different conditions.

2.4. The brewery pipe-line (failure analysis)

All good drinking polymer technologists would be very upset to hear that 10 000 litres of their favourite brew had been lost due to pipe failure and would demand that the reason for failure was determined and those responsible punished.

Failure analysis is, to a large extent, detective work. The important feature of the test methods used is that they should discriminate between good and bad, and it is of little importance if they are non-standard nor whether their absolute accuracy is high.

However, if a specification for the material exists the tests therein can be used as a first step to show whether or not there had been a quality-control lapse. Similarly, if there is a product specification it may be used as a basis for showing whether or not there were manufacturing faults. This type of evidence can be particularly valuable when legal action follows a failure.

If use of these tests fails to find a fault and there was no evidence of deliberate or accidental damage, nor of unreasonable use, it would imply that the performance specification tests were inadequate. Such cases have occurred with pipes and hoses, and indicate a need for a better understanding of service conditions and for a test which is better related to those conditions.

2.5. Further consideration of the examples

In taking my examples I have cheated somewhat by using reasonably simple cases and only dealing with one aspect of, or rather reason for, testing in each case. Furthermore, I resolutely avoided getting to details where inevitably many more problems would crawl out from under the stones.

As illustration of this, it is doubtful if very reliable scratch and abrasion tests exist to prove the fitness of a bath; one would not want the expense of designing even a bath by trial and error, and design data would be subject in principle to the same problems as for a boat—it would need to be proved that rapid quality-control tests on the material of a bath were capable of relating to a faulty product. I doubt if the tests in the present Standards (as opposed to the tests on the bath) are 100% satisfactory.

3.   REQUIREMENTS FOR TESTING

Looking again at the reasons for testing, it is possible to link them to the role of Standards:

Quality control—a quality-control specification and a quality-control schedule.
Design data—possibly codes of practice for generating and for using data.
Predicting service performance—a performance specification.
Investigating failures—the QA and performance specifications can be used.

It is interesting to note that Standards in one form or another can play a role in all the reasons for testing and are not purely concerned with quality assurance. There are different types of Standard and they each have a particular function.

Clearly this raises possible problems by asking whether our current Standards are adequate. Probably a great many are not, particularly performance specifications which are really quality-control methods, and the absence of many good design Codes of Practice.

There are a number of general requirements for a test method; it must have adequate precision, reproducibility, and so on. There are, however, particular attributes related to the reason for testing.

1. For quality control: the test should preferably be as simple, rapid and inexpensive as possible. Non-destructive methods and automation may be particularly attractive. The best tests will additionally relate accurately to product performance.

2. For producing design data: the need is for tests which give material properties in such a form that they can be applied with confidence to a variety of configurations. This implies very considerable understanding of the way material properties vary with geometry, time, etc. Extreme speed and cheapness are of relatively minor importance, there is little interest in non-destructive methods. For complex and long-running tests, automation may be desirable.

3. For predicting service performance: the essence of the test must be that it relates to service—the more relevant the test to service conditions, the more satisfactory it is likely to be. Extreme speed and cheapness are less likely to be important but there is a need for test routines which are not excessively complex. Non-destructive methods may be acceptable.

4. For investigating failures: at least half of the battle is knowing what to look for, to prove that it needs, more than anything, a test which discriminates well. There is often no need for absolute accuracy or in some cases even relevance to service.

There is of course nothing black and white about attributing these requirements to the reasons for testing, but they indicate the emphasis which usually applies in each case.

4.   GROUPING OF TESTS

It may be useful to consider how we group or classify the various parameters to be measured and the methods used. Most commonly we come across groupings such as:

Mechanical
Thermal
Rheological
Dimensional
Electrical
Chemical
Surface
Environmental, etc.

Such groupings are very convenient for text-books and the headings are necessary to indicate which types of property we are talking about. However, it is not very helpful when we consider the reasons for testing and what we need to get from a test.

We can sub-divide, for example mechanical tests, and list the actual properties we have interest in: strength, stiffness, creep, and so on. This is necessary to indicate which parameter we need to measure, but again it is not very helpful in terms of why we are testing and which method we should use.

One way of classifying tests might be to think of:

Fundamental properties or tests
Apparent properties or tests
Functional properties or tests

Whichever type of propery and whatever parameter we choose, this classification can help in considering what is needed from the result and hence which test method should be used.

Taking the example of strength, the fundamental strength of a material is that measured in such a way that the result can be reduced to a form independent of test conditions. The apparent strength of the material is that obtained by a standard method which has completely arbitrary conditions. The functional strength is that measured under the mechanical conditions of service, probably on the complete product.

This classification can be loosely related to the reasons for testing and what we need to get from a test. For quality control we do not need to know fundamental properties, apparent properties will often be acceptable, with functional properties certainly being desirable. For design data we really need fundamental properties although we can get considerable help from functional properties. For predicting service performance, the most suitable properties would be the functional ones. For investigating failures we are again unlikely to need fundamental methods.

It is worth noting that when you look at properties or tests in this way the gaps become clear: for example, most existing measures of impact strength or melt flow yield apparent properties and there is a need for fundamental methods, whereas most dimensional, and many thermal and chemical tests give fundamental properties.

It is also worth noting that if we are measuring the effects of environment, weathering for example, for use as design data, it may not be necessary to use a method giving absolute results to monitor the changes with time, an apparent method may suffice.

5. THE PROBLEMS

The factors which have been discussed have been collated in Table 1, not because I have any delusions that they form a neat or perfect model for testing but as a vehicle to help identify what are some of the problems or areas for development in testing.

TABLE 1

Reason for Testing	Standards	Testing needs	Property class
Quality control	QA specification QA schedule	Simple, rapid Relates to performance	Apparent Functional
Design data	Codes of practice	Results as function of shape, time, etc.	Fundamental Functional
Predicting service performance	Performance specification	Relates to service	Functional
Investigating failures	QA specification Performance specification	Discriminating power	Apparent Functional

A number of problems can be traced to the present state of Standards. There is too little appreciation of the different types of Standard and so often wrong or inadequate tests are used in them. In some cases even quality-assurance specifications get contaminated with bits of performance specifications. Very often, so-called performance specifications are no more than glorified quality-assurance specifications.

The building blocks of a specification are test method standards and these may attempt to cater for the quality controller and the designer, and please neither.

There are, of course, far too few Codes of Practice for design. This cannot be due only to a lack of knowledge but must also be due to a lack of interest and effort.

In formulating the test methods that we use, I would like to see more understanding and cooperation between the different testing interests, as for example, used earlier, the university researcher and the factory floor quality controller.

It is perhaps in the testing needs that we would expect to identify the most problems or lack of adequate methods.

There will always be a need to improve the cost and precision efficiency of quality-control tests, and modern instrumentation technology is being applied to this end. Perhaps more important, and more difficult, is the need for efficient tests which also relate better to performance. These two requirements have a nasty habit of conflicting when designing a test.

Very many standardised tests have little value for design data purposes as they stand, because of the completely arbitrary conditions specified. There are still many gaps in our knowledge as regards the rules to relate materials tests to complex shapes, dynamic conditions, long times and to incorporate the effects of environment—the problems identified in the example of the GRP boat.

We need what I have termed 'fundamental tests' for design purposes and for many properties these are lacking. Similarly, for performance specifications we need functional tests which relate to service and again these are often lacking.

In the areas where there are gaps, there is clearly a need for the development of suitable tests and for research to understand material behaviour such that designs and extrapolation rules can be formulated. However, I also suspect that there are many instances where reasonable tests and rules exist but are not always used and have not been publicised in suitable Codes of Practice. Furthermore, whilst it is right and worthy to strive for better tests and understanding, we should also learn to make the most of what we already have.

For the papers on particular areas of testing which follow, I would ask, 'Why are you developing your test?', 'What do you need to get from the results?', and 'What effort are you going to make to see that the information is used in real practical situations?'.

Polymer Testing **4** (1984) 101–115

Thermal Analysis of Polymers using Quantitative Differential Scanning Calorimetry

M. J. Richardson

National Physical Laboratory, Teddington, Middlesex TW11 0LW, UK

SUMMARY

A brief general description of differential scanning calorimetry is followed by a more detailed treatment of the quantitative aspects of three applications that have widespread usage in the polymer industry: (1) the glass transition; (2) heats of reaction (curing); and (3) crystallinity.

Emphasis is placed on the true meaning of the derivable physical quantities and it is shown that an uncritical approach can result in errors that are far beyond the instrumental uncertainties. By contrast, correct data treatment allows the rapid and accurate characterisation even of materials such as glasses which have long been regarded as having little to offer but qualitative generalisations.

1. INTRODUCTION

Differential scanning calorimetry (DSC) is perhaps the best-known example of the growing number of techniques which come under the broad heading of 'thermal analysis', a generalisation for the measurement of some material response to a programmed change of temperature. The parameter monitored may range from the familiar, such as a mass or a dimension, to some less common optical or acoustical property. DSC measures heat capacities (c_p) which were only accessible to a very few laboratories when the technique was first introduced

101

in the late 1960s. Such has been the subsequent growth that c_p can now often be determined more easily than the density of a material. This facility is especially valuable for polymers because processing, or any mechanical and/or thermal treatment, confers its own 'fingerprint' on the arrangement of polymer chains in the resultant product. The energy of the system is accordingly unique and the change, the integral of c_p, to a well-defined reference state (commonly the molten polymer) immediately characterises the prior history of the sample.

In DSC the *directly measured* quantity is either a differential power (ΔW) or temperature (ΔT) between two cells (Fig. 1), one containing the sample and the other a thermally inert reference, transformation to c_p readily follows (see below). In a differential power instrument (broken line, Fig. 1) the two cells have individual heaters and ΔW is the difference between the power output of these that is needed to maintain the same programmed temperature in each cell. At a melting point (T_m), for example, the sample heater output must increase considerably (Fig. 2) as heat of fusion is supplied. In a differential temperature instrument (dotted line, Fig. 1) both cells respond to a common power source and the signal ΔT arises from the differing thermal paths to, and properties of, the two cells. At T_m the sample temperature remains constant until all heat of fusion has been supplied, no such constraint applies to the reference cell so that $\Delta T = T_s - T_R$ becomes very negative in this region. The abscissa of Fig. 2 is shown as time or temperature and with modern instruments the two are virtually synonymous apart from minor differences at large thermal events.

The signal (S) from either type of DSC is proportional to the difference in heat capacity between sample and reference states.

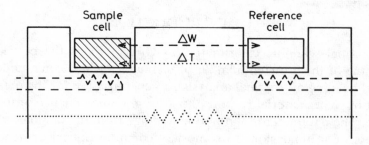

Fig. 1. Schematic representation of differential power (−−−−) and differential temperature (. . . .) DSC.

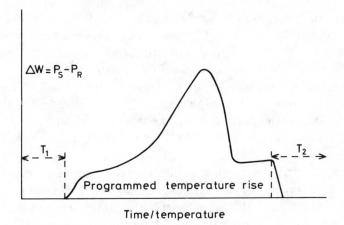

Fig. 2. The differential power signal through a phase change. For differential temperature instruments the signal is similar but of opposite sign.

Quantitative operation calls for three experiments with the sample cell containing: (1) empty pan (e); (2) pan+sample (s); (3) pan+ calibrant (c). The reference cell is undisturbed throughout, it usually contains an empty pan to roughly balance the effect of the pan in the sample cell. If ΔQ is the difference in heat capacity between the 'empty' cells of case (1) the signal is given by

$$\text{Case (1)} \quad S_e = K \Delta Q \tag{1}$$

$$\text{Case (2)} \quad S_s = K(\Delta Q + m_s c_{ps}) \tag{2}$$

$$\text{Case (3)} \quad S_c = K(\Delta Q + m_c c_{pc}) \tag{3}$$

and the basic equation of quantitative DSC emerges as

$$c_{ps} = \frac{(S_s - S_e) m_c c_{pc}}{(S_c - S_e) m_s} \tag{4}$$

where m and c_p refer to mass and specific heat capacity, respectively. α-Alumina is widely used as the calibration material. Various precautions are needed to ensure the correct application of eqn (4) but these have been fully described[1,2] and it is the intention in this paper to concentrate on the use and abuse of the resultant data. This is especially relevant at the present time when the derivation of c_p is rendered ever simpler by the increasing use of computerised methods of data treatment. These are also very useful for the numerical integration of the c_p-temperature curve to give the enthalpy changes that are frequently mentioned in this paper. The methods previously used, planimetry or weighing cut-out areas, inevitably introduce errors over and above those of the DSC experiment itself.

In the following sections several common applications of DSC will be considered with special emphasis on the true meaning of the data generated. Experimental details are kept to a minimum; results were obtained using a Perkin Elmer DSC2 and data treatment as described in References 1 and 2.

In general, results reproduce values found by adiabatic calorimetry to ±1%. Measurements can be made in heating or cooling and it is good practice to check the agreement between the two modes—in the liquid state, for example, c_p should be independent of both the magnitude and sign of the programmed temperature change as well as the sample size and geometry.

2. THE GLASS TEMPERATURE AND CURE OF AN EPOXY COATING MATERIAL

2.1. General

Figure 3 shows the DSC record of the cure behaviour of a typical epoxy pipeline coating material as it is heated at 20 K min^{-1}. Various zones (A, B, etc.) are indicated and a clear understanding of their significance is essential for any subsequent treatment of the glass transition (T_g) or the heat of reaction (ΔH(reaction)). No attempt is made to discuss the very complex reaction chemistry, for present purposes it is sufficient to note that polymerisation takes place with no elimination of low molecular weight species so that the weight of both reactants and product is the same.

Reactants are compounded to give a glassy powder (A) which liquifies at the glass transition (B). The range of purely liquid behaviour (C) is very restricted because the exothermic curing reaction (D) soon depresses the apparent heat capacity. At the end of the reaction c_p is that of a rubbery cross-linked material (E), which has a small positive temperature coefficient ($1\cdot46 \times 10^{-3}\,\mathrm{J\,g^{-1}\,deg^{-2}}$). On cooling to room temperature and reheating a much simplified curve is obtained. The only 'event' is the glass transition (G) at about 370 K

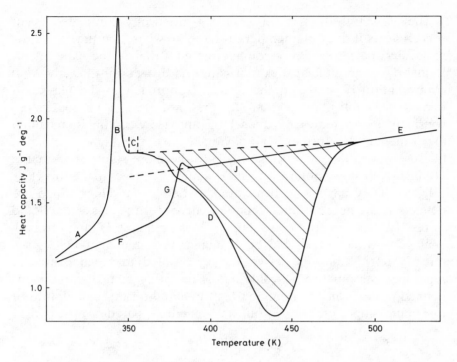

Fig. 3. The heat capacity and curing exotherm of an epoxy powder (A, B, etc.) and the resultant coating (F, G, etc.) as they are heated at 20 K min^{-1}. The lettered zones are identified in the text. The shaded area corresponds to an apparent heat of reaction of 54·5 J g^{-1}, the true value at 350 K is 44·7 J g^{-1}.

where the low-temperature glass (F) transforms to a rubber (J), the curve for the latter merging smoothly with the high-temperature portion (E) of the previous run.

2.2. The glass transition (T_g)

Although the T_g regions shown in Fig. 3 (B, G) are superficially very different for reactants and products, a common feature, an abrupt increase in c_p, emerges if the dissimilar peak magnitudes are temporarily ignored. This step-like increment, Δc_p, is characteristic of the glass transition region of any material and is associated with the onset of translational degrees of freedom either of whole molecules or of lengthy chain segments.

The glass transition is very much a kinetically dominated event reflecting, as it does, the temperature region where the time-scale for molecular motion becomes comparable with that of the experiment. Below T_g some molecular conformation is 'frozen in'—exactly *what* conformation depends on the route of approach to the glass. An immediate implication is that T_g does not have a unique value; consider the extremes of quenching and slow-cooling (annealing) experiments, the former will obviously give the higher T_g value because relaxation cannot take place to the more stable state that is attainable on slow cooling. Specific volume, V, curves make this clear (Fig. 4). They have long been used to derive T_g values which are defined as the point of intersection of curves for the glassy and liquid states. The calorimetric equivalent of V is enthalpy, H, and formally analogous values of T_g can be obtained,. The fact that DSC gives directly not H but its derivative, c_p, has led to much confusion regarding calorimetric T_g values. It is tempting to take some well-defined point from the c_p region (extrapolated onset, temperature of maximum slope, etc.) as T_g but an immediate paradox arises when

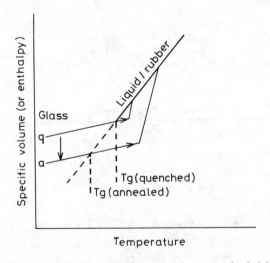

Fig. 4. Specific volume (or enthalpy) curves for quenched (q) and annealed (a) glasses. The latter may be obtained by slow cooling or by allowing the quenched material to relax (arrowed path) at temperatures slightly below T_g. On reheating the 'q' form superheats much less than 'a' form and the apparent glass transition temperatures (measured from the discontinuity in the DSC curve) are reversed.

considering quenched and annealed samples—the anticipated order is reversed. Again the problem has a kinetic source and the paradox vanishes at very slow heating rates, the use of which, however, negates one very important practical advantage of DSC—rapid experimentation. Fortunately by following a very simple procedure DSC measurements will give T_g values that truly characterise the state of the low-temperature glass and are independent of instrumental heating rate. The peaks of Fig. 3 reflect the tendency for annealed glasses to superheat at typical DSC scanning rates. Again this is shown schematically in Fig. 4 from which it is clear that the DSC peak represents the enthalpy increment needed for the material to 'catch up' with the equilibrium liquid line. If two temperatures T_1 and T_2 are chosen that are unambiguously in the glassy and liquid regions, respectively, it is a straightforward procedure to integrate the DSC-generated c_p–T curve to give $H_1(T_2) - H_g(T_1) \equiv Z$. Heat capacities below T_1 and above T_2 can be represented by linear equations which on integation give

$$H_g(T) = aT + 0\cdot5bT^2 + c \tag{5}$$

$$H_1(T) = AT + 0\cdot5BT^2 + C \tag{6}$$

substituting T_1 in eqn (5) and T_2 in eqn (6) followed by subtraction gives the integration constants C–c in terms of Z and other explicit quantities. Knowing C–c, T_g is calculated as a solution of the equation $H_g(T_g) = H_1(T_g)$. This computational approach is very successful and for any polymer (or, indeed, any organic glass) a given thermal history yields a T_g value that is reproducible to ±1 K. The computational aspect is emphasised because the slopes of the enthalpy curves (i.e. c_{pg} and c_{pl}) differ only slightly (about 20% for polystyrene) and visual inspection of H–T curves is not quantitatively successful. By contrast, the expansion coefficient for this polymer more than doubles on passing through T_g and it is fairly easy to observe the intersection temperature of specific volume curves. Dilatometry, however, has inherent sampling limitations (size, geometry, voids) which do not influence correctly designed DSC experiments and the calorimetric technique is much to be preferred for routine measurements.

Regions B and G of Fig. 3 can now be re-examined in the light of the above discussion. Although annealing has so far implied slow cooling through T_g, isothermal holding at temperatures in the region

between about T_g–30 and T_g is also effective and the very large
superheating peak at B no doubt reflects annealing at high storage
temperatures for this particular sample which was from a batch that
had been used in a desert region of Australia. Had it been preheated
to 350 K before cooling and running as in Fig. 3 only a small peak, of
a size comparable to that in region G, would have been found.

Sample pretreatment is essential in one important contemporary
use of this region—monitoring the state of cure of a pipeline coating
(Fig. 5). In principle this is a simple procedure; the glass temperature
is measured before and after exposure to a temperature that is high
enough (530 K for the example shown) to allow any residual cure to

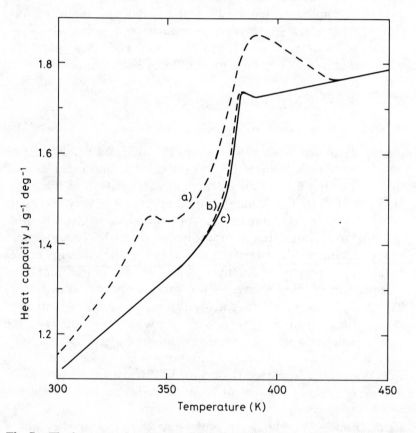

Fig. 5. The heat capacity of a cured epoxy pipeline coating: (a) as received; (b) after
cooling at 40 K min^{-1} from 430 K; and (c) as (b) but cooled from 530 K. T_g is about
1 K higher than after treatment (b). All heating rates are 20 K min^{-1}.

take place. Both measurements must refer to identical glass formation conditions but those for the sample as received ((a), Fig. 5) are usually unknown and may include additional perturbation by internal stress or even plasticisation by moisture. The beneficial effects of thermal pretreatment are clearly shown by the transformation to the comparable curves (b) and (c) of Fig. 5.

Although it is an oversimplification to use a single value of T_g to describe the result of a particular glass formation process it is a fact that DSC is one of the few simple and easily applicable techniques that can be applied to a process that has wider implications for the long-term properties of amorphous polymers than is generally recognised. The reduction of impact strength on physical ageing, for example, is essentially the result of an annealing process that is very conveniently monitored by DSC.

The value of T_g for polymers generally decreases by 2–3 degrees per decade decrease in cooling rate. Most polymers are fairly rapidly cooled, whether as pellets or moulded products, and determinations of T_g are therefore made on samples with roughly comparable histories. Large discrepancies can emerge if storage temperatures are such that isothermal annealing is possible (B, Fig. 3), time then becomes important and polyvinyl acetate, for example (with T_g only a little above ambient) can show a wide range of values.

2.3. Heat of reaction

The onset and termination of cure at 350 and 490 K, respectively, are readily identifiable in Fig. 3 and the overall enthalpy change is

$$H_1(490, \text{product}) - H_1(350, \text{reactants}) = 201 \cdot 1 \text{ J g}^{-1}$$

The enthalpy of reaction may be referred to any temperature if the heat capacities of the reactants and products are known. In fact, from Fig. 3, the former can only be obtained in the vicinity of 350 K whereas the latter exist from 390 to 520 K, and it is obviously sensible to make the relatively short extrapolation to 350 K to give

$$H_1(490, \text{product}) - H_1(350, \text{product}) = 245 \cdot 8 \text{ J g}^{-1}$$

from which it immediately follows that

$$\Delta H(350, \text{reaction}) = H_1(350, \text{product}) - H_1(350, \text{reactants})$$
$$= -44 \cdot 7 \text{ J g}^{-1}$$

This is unambiguous and refers to a thermodynamically well-defined state. Although data for the molten reactants are not available above 350 K they can often be obtained via subsidiary experiments on the individual components and additivity tested by comparison with the direct measurement at 350 K.

Many manufacturers now offer data treatment facilities which in the case of Fig. 3 could be used to produce a baseline linking 350 and 490 K (broken line), the area defined by this and the cure exotherm is then taken to represent the heat of reaction. From the data of Fig. 3 the baseline corresponds to an enthalpy change of $255 \cdot 6$ J g^{-1} whence 'ΔH(reaction)' $= -54 \cdot 5$ J g^{-1}—more than 20% greater than the above value at 350 K. In fairness it should be observed that the heat of reaction at any temperature above 350 K will increase with any reasonable assumption about the temperature dependence of c_{pl} (reactants) (e.g. that it parallels that of the products) so the discrepancy will decrease at higher temperatures. The example does, however, highlight the dangers inherent in a 'black box' approach to computational aids—users must constantly ask themselves what is the significance of the data that they are generating. The difficulties are perhaps most acute when trying to extract isothermal reaction data from non-isothermal exotherms. The problem is a difficult one at the best of times and in some programs the solution is made impossible by the use of unsound thermochemical foundations on which to base the subsequent complex computational structure.

3. MELTING AND CRYSTALLISATION OF POLYMERS

Some of the earliest applications of DSC were to crystalline polymers because it was intuitively obvious that the area of the melting peak would increase as the degree of crystallinity (x) increased. This generalisation, although true, concealed a number of problems which are rarely considered even now and as a consequence data often have little more than qualitative significance. The difficulties are best illustrated by reference to a real example; Fig. 6 shows results for drawn 6/12 nylon filaments as they are cycled at ± 20 K min^{-1} over the temperature range from 340 to 530 K. The curves are complex and raise (more so than if the example chosen has been polyethylene)

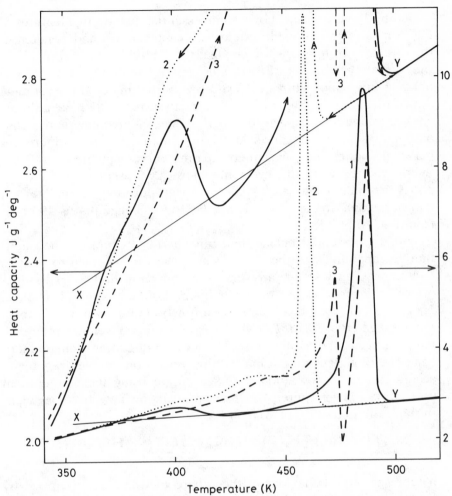

Fig. 6. The heat capacity of 6/12 nylon showing: 1, drawn filaments (——); 2, crystallisation from the melt (. . . .); and 3, remelting (– – –). The fine-scale curves (left-hand ordinate) show the low-temperature thermal events in more detail. See text for discussion of the baseline XY. All rates are ±20 K min⁻¹.

a number of questions:

1. What construction will give the heat of fusion (ΔH)?
2. How are multiple peaks accommodated?
3. To what temperature does ΔH refer?
4. Can ΔH be related to the room temperaure crystallinity to which most other measurements refer?

Before discussing the above, some essential quantitative features of Fig. 6 must be mentioned. Liquid heat capacities are independent of the direction of temperature change, the fine-scale portion of Fig. 6 shows that the span of all values covers less than 1% and this also holds at low temperatures for DSC-crystallised material (the original drawn filaments, having a different structure, naturally have a different crystalline heat capacity). Superheating, supercooling and secondary recrystallisation effects imply that there are great differences between cooling and subsequent heating curves but the overall enthalpy change over the cycle 500–340–500 K should be zero (at 340 K crystallisation is, for all practical purposes, complete). A typical residual for this cycle is 3–4 J g^{-1} in a change that is 490 J g^{-1} from 340 to 500 K.

With this essential background information to support the quantitative capabilities of the DSC technique we can return to queries 1–4 above. Fairly simple answers can be given if the DSC is thought of not in terms of c_p but rather its integral, an enthalpy (H) change. $H_l(T_R) - H_x(T)$ may be obtained as a function of T over any temperature range from T to T_R, where T_R is some convenient reference temperature in the molten state. Numerical integration using modern computational techniques makes this operation simple, rapid and accurate; it is slower and far less accurate using more traditional procedures. The required baseline construction now becomes clear through the definition of $\Delta H(T)$

$$\Delta H(T) = H_l(T) - H_x(T) = [H_l(T_R) - H_x(T)] - [H_l(T_R) - H_l(T)]$$

$$(7)$$

it represents the behaviour of the supercooled liquid as required by the second square-bracketed term in eqn (7). On a DSC curve this corresponds to the low-temperature extrapolation of the liquid heat capacity (XY, Fig. 6)—a fairly straightforward procedure because c_{pl} is almost invariably a linear function of temperature. Of course, for high-melting polymers the extrapolation to room temperature (see below) is a searching test of the accuracy of $\partial c_{pl}/\partial T$ and it is only prudent to maximise the liquid range as much as possible, for example by cooling experiments (most polymers supercool several tens of degrees) or by quenching to the fully amorphous state (when the sample between T_g and the onset of crystallisation represents highly supercooled liquid). The simple XY baseline construction of

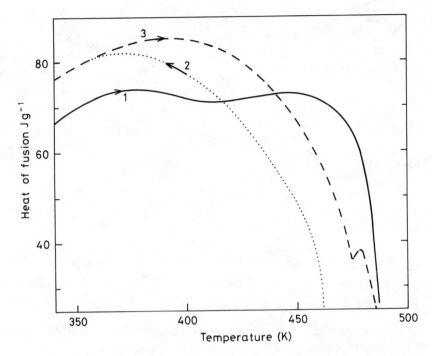

Fig. 7. The heat of fusion of 6/12 nylon. Curves are as identified in Fig. 6.

Fig. 6 is the route to successful answers to all of the queries 1–4 above. The presence of several peaks or a generally complex background is accommodated quite naturally and the resultant ΔH curves (Fig. 7) refer to a set of well-defined values of T. The final query, 4 above, remains to be answered and must be preceded by a brief discussion of crystallinity–temperature curves.

When a crystalline polymer is cooled, x increases until it reaches some low-temperature limit x_c (Fig. 8) which is a function of crystallisation conditions. For two samples with different crystallisation histories the ratio x_1/x_2 will change with temperature until both are in their x_c regions and $x_{c1}/x_{c2} \neq f(T)$ (it would be a remarkable coincidence if at higher temperatures, where $x < x_c$, both x_1 and x_2 changed so as to maintain a constant ratio). Enthalpic crystallinities are defined by

$$x(T) = \Delta H(T)/\Delta H_\infty(T) \tag{8}$$

where ΔH_∞ refers to the perfectly crystalline polymer. The onset of

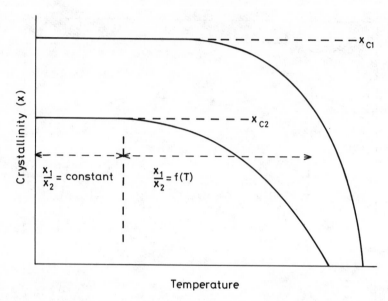

Fig. 8. The variation of crystallinity with temperature.

melting in one of the samples can therefore be found by observing at what temperature the ratio $\Delta H_1(T)/\Delta H_2(T)$ ceases to be constant. In this way it is possible to relate high-temperature DSC measurements to more conventional procedures which generally refer to results obtained at room temperature. It is also possible to start DSC runs well below room temperature for a direct measurement of ΔH at this point, and in this way it can be shown that low-density polyethylene, for example, is already 'melting' at ambient temperatures. Relative crystallinities are readily obtainable using DSC results and the overall accuracy can be high if measurements are related to some internal standard such as material crystallised in the DSC. This is because in the sequence of operations to give data on (1) as received (2) DSC-crystallised material, systematic errors (e.g. in sample location) are of the same sign and tend to cancel out.

Absolute values of calorimetric crystallinities require $\Delta H_\infty(T)$ and this is usually obtained by inserting an independently determined value of x (using x-rays, for example) into eqn (8). This procedure assumes that the two techniques have identical responses to a whole range of often very imperfect structures and the assumption is almost certainly incorrect in detail. It is sufficient to observe that agreement to within a few percent is general at present and will improve as our detailed knowledge of polymer structure increases.

We conclude this section with some comments on the behaviour of the 6/12 nylon of Figs 6 and 7 in the light of the above discussion. Structural changes (perhaps relaxation to a lower draw ratio) in the original filaments are obvious at temperatures nearly 100 K below the eventual melting region. Any data referring only to higher temperatures will therefore be meaningless so far as the original material is concerned and measurements must refer to 360 K, at most, if the structure of the 'as received' material is to be correctly defined. Equation (8) is based on a simple additivity concept for contributions from crystalline and amorphous regions. It has already been noted that structural imperfections (both surface and bulk) will perturb this and an additional complication for drawn material is the effect of orientation on the amorphous contribution—again this will affect absolute values of x as determined by different techniques but a purely calorimetric figure still has great value for the purpose of material characterisation.

The complex shape of the melting curve of 6/12 nylon crystallised in the DSC is basically due to a phase change at high temperatures. The original structure starts to melt at about 470 K but transforms to another form before melting at about 490 K. Detailed invstigation of the two forms is possible by varying the heating rate and using high-temperature isothermal annealing.

Figure 7 shows the heat of fusion of the 6/12 nylons of Fig. 6. Immediately obvious is the importance of specifying the temperature to which a given measurement refers. The overall shapes of the curves of Fig. 7 are due to two effects: (1) in general, $c_{px} \neq c_{pl}$ and so it is a physical necessity that $\Delta H(T) = f(T)$; and (2) at higher temperatures when $x < x_c$, $\Delta H(T)$ falls off due to a reduction of crystallinity. For the samples of Figs 6 and 7 the enthalpy ratio is constant at 0.887 ± 0.002 over the range 340 to 370 K and it is safe to conclude that this will also represent the ratio of room temperature crystallinities.

REFERENCES

1. Mills, K. C. and Richardson, M. J. (1973). *Thermochim. Acta*, **6,** 427.
2. Richardson, M. J. (1978). In: *Developments in Polymer Characterisation—1*, Ed. by J. V. Dawkins, London, Applied Science Publishers Ltd, Chapter 7.

Polymer Testing **4** (1984) 117–129

Dynamic Mechanical Method in the Characterisation of Solid Polymers

R. E. Wetton

Department of Chemistry, Loughborough University of Technology, Loughborough, Leicestershire LE11 3TU, UK

SUMMARY

The basic theory behind dynamic mechanical measurements is explained and dynamic moduli defined. The effect of changing the major variables temperature and frequency in a region of polymer relaxation is discussed.

Commercially available instrumentation is briefly reviewed with the merits of the various systems explained.

Measurements on random and block copolymers are compared and the power of the technique for distinguishing phase separation and partial compatibility is emphasised. Crystalline polymers present interesting two-phase composites, with the crystalline phase and related structures contributing specific relaxations particularly in the region below the melting point.

Carbon-reinforced epoxies represent more traditional composites where the engineering data obtained directly from the measurements are as important as the relaxational characteristics. The power of frequency multiplexing in fully characterising molecular relaxation is illustrated with these materials and activation energies derived from a single thermal scan.

1. INTRODUCTION

Solid polymers and viscoelastic liquids can be extensively characterised by the dynamic mechanical technique which involves applying an

117

Polymer Testing 0142-9418/84/$03·00 © Elsevier Applied Science Publishers Ltd, England, 1984. Printed in Northern Ireland

oscillating mechanical strain and resolving the stress into real and imaginary components. This procedure essentially detects all changes in the state of molecular motion as temperature is scanned. It is thus a most powerful technique for studying the effect of not only molecular structure but also phase morphology and filler addition on the physical properties required for component design.

The torsion pendulum technique in particular has been used extensively to investigate grain boundary and crystal dislocation motion in metals.[1] In the polymer field, Alexandrov and Lazurkin[2] measured the frequency dependence of the moduli of rubbers as early as 1940. The first workers to use the technique for the dynamic mechanical thermal analysis of polymers were Schmieder and Wolf,[3] using a torsion pendulum device.

The advent of microprocessor control of instrumentation has led to the availability of commercial systems which are as simple to operate as DTA instruments. A brief review of instruments designed principally for thermal scanning work, such as the Polymer Laboratories' DMTA and the DuPont DMA together with more massive instruments with which slow temperature scans can be achieved, is given in the experimental section.

The dynamic mechanical technique is largely complementary to DSC/DTA. These latter techniques give quantitative measurements of heat changes during first-order thermodynamic transitions (e.g. melting and crystallisation). Resolution of pseudo second-order transitions (T_g) is rather poor particularly in the case of minor components, and detection of secondary transitions is almost impossible.[4] The dynamic mechanical method detects molecular relaxations such as T_g with a factor of approximately 1000 higher sensitivity than DSC/DTA and measures second-order transitions quantitatively. The response to first-order transitions is, however, complex and at the moment poorly investigated. As will be seen later, there are characteristic relaxations occurring in this region.

Examples are given of applications to phase-separated copolymers, rubber toughening, filler reinforcement and crystallinity effects in thermoplastics, epoxy and silicone rubber cross-linking studies, the study of coatings and films and finally, using frequency multiplexing, the rapid generation of relaxation spectra and activation energies.

2. THEORY

When a sinusoidal stress is applied to a perfectly elastic solid the deformation, and hence the strain, occurs exactly in-phase with the stress. In extension or bending after allowance for the correct geometrical factors the dynamic Young's modulus (E^*) is given basically as stress amplitude/strain amplitude. In shear deformation the dynamic rigidity modulus (G^*) is obtained. However, when some internal molecular motion is occurring in the same frequency range as the impressed stress, the material responds in a visco-elastic manner and the strain response lags behind the stress.

It is then convenient to resolve E^* or G^* into perfectly elastic and perfectly viscous components, called the storage (E', G') and loss components (E'', G''), as shown in the Argand diagram in Fig. 1. A

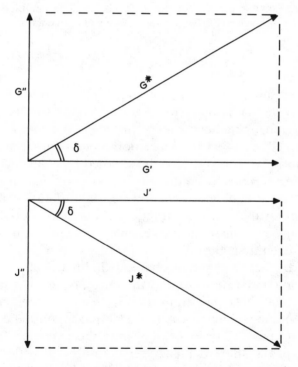

Fig. 1. Argand diagram showing resolution of complex moduli (G^*) or compliances (J^*) into their storage and loss components.

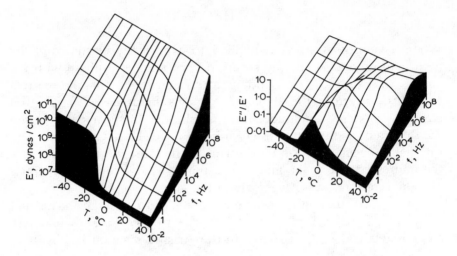

Fig. 2. Storage modulus and damping for SBR rubber (Nolle[5]) in three-dimensional representation showing clearly the temperature and frequency plane effects.

more useful parameter is the dimensionless ratio, $\tan \delta = E''/E'$, often referred to as the damping factor. Figure 2 shows the classic data of Nolle[5] for styrene–butadiene rubber through the main T_g relaxation region. It can be seen that the relaxation process (onset of some new degree of motional freedom) may be scanned by changing frequency at constant temperature (frequency plane data) or changing temperature at constant frequency (temperature plane data). Although the former are purer data, in that structural changes may occur during a thermal scan, experimentally accessible frequency ranges are so limited that even a single relaxation process cannot conveniently be encompassed. The 'dynamic mechanical thermal analysis' technique is thus to scan temperature over a wide range, typically from $-150\,°C$ to above the materials glass transition. In this way secondary processes (β and γ) due to limited chain or side-group motion can be seen as well as the main T_g processes (α) in a 'thermal spectrum'.

Damping peaks shift to higher temperatures T with higher impressed measurement frequencies f(Hz). The shift allows the activation energy for the process to be determined as $\Delta E = -R[\mathrm{d} \ln f/\mathrm{d}(1/T)]$. Further background theory can be found in References 6 and 7.

3. INSTRUMENTATION

After a number of decades of work with home-made equipment there are now a number of commercial instruments on the market. The original Rheovibron is still best for fibres although very laborious in the standard version. For solid polymers the PL-DMTA has the best dynamic ranges and a useful frequency range of three and a half decades. The DuPont DMA and the Brabender–Lonza torsion pendulum have good sensitivity but suffer from drifting frequency because they measure at sample resonance and can only handle relatively stiff samples. The Rheometrics range offers the facility of continuous shear in liquids as well as giving some frequency coverage with somewhat limited dynamic range. The Dynastat (IMASS) instrument has the ability to work at relatively high static loads with superposed sinusoidal deformation.

The PL–DMTA is shown schematically in Fig. 3. In this instrument, stress is proportional to the level of a.c. current fed to the drive

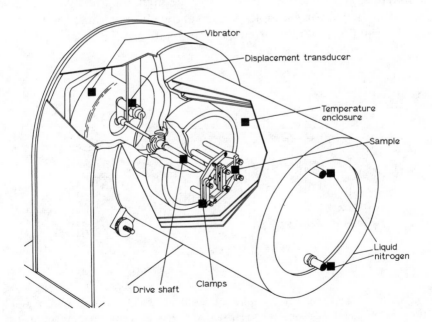

Fig. 3. Mechanical head of PL–DMTA showing the essential features of sample mounting, vibrator system and transducer.

coil from the analyser module. The frequency of oscillation is selectable from 0·033 to 90 Hz and does not depend on sample stiffness. Strain is proportional to the displacement of the drive clamp and is monitored by a non-contacting eddy current transducer. The analyser unit compares the stress and strain signals and by using refined counting circuits resolves the strain into its in-phase (storage) and out-of-phase (loss) components. When the sample geometry constant is dialled into the instrument $\log E'$ and $\tan \delta$ are computed. The sample temperature is controlled in the range $-150\,°C$ to $+300\,°C$ by a temperature programmer such that it can be ramped up or down at controlled rates (up to $15\,°C\,min^{-1}$ for measurement) or isothermed at any desired temperature. Use of a microcomputer and the IEEE interface allows frequency to be multiplexed during a slow thermal scan and with all data stored for subsequent manipulation.

The normal mode of deformation geometry is by bending small bars as dual or single cantilevers as shown in Fig. 4(a). In the single cantilever mode considerable thermal expansion, such as occurs through a melting point, can be accommodated because of the lateral compliance of the drive. Alternative geometries are available and Fig. 4(b) shows, for example, shear sandwich geometry, for measuring the rigidity modulus of rubbers and soft adhesives.

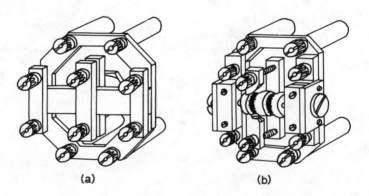

(a)   (b)

Fig. 4. (a) Dual cantilever clamping of small rectangular bar sample, typically $30 \times 10 \times 2$ mm in dimensions. (b) Shear sandwich geometry for clamping soft materials such as adhesives and rubbers. Horizontal and vertical orientations can be employed if required.

4. RESULTS AND DISCUSSION

4.1. Phase structure measurements

In random copolymers two inherently incompatible polymer sequences are forced to co-exist in a single phase. The polymer exhibits a
single relaxation process intermediate between that of the parent
homopolymers. Figure 5 illustrates this for poly(ethylene-ran-vinyl
acetate) and two different frequencies. Although sharpening of loss

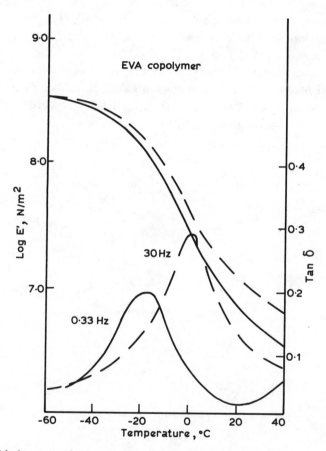

Fig. 5. Ethylene-ran-vinyl acetate copolymer at two different frequencies on the
Polymer Laboratories DMTA. A single-phase polymer exhibits a single T_g relaxation
between that of the parent polymers.

peaks of random copolymers can occur because of improved compatibility between sequences, it is thought that changes in ethylene sequence crystallinity are the cause in this case. In contrast, incompatible sequences in block copolymers, in general, phase separate. This can be assessed by DMTA via the observation of two separate processes. Figure 6 shows this for effectively a block copolymer of styrene–butadiene–styrene random sequences. However, the poly(styrene) peak in particular is occurring at a lower temperature than for homopoly(styrene) indicating some mutual solution and phase boundary mixing. The modulus level in the region between the two T_g processes allows an assessment of the phase structure. If the glassy phase is continuous the modulus will be high but if the rubber phase is continuous it will be low. These effects can be quantified by model studies.

Rubber toughening of glassy polymers such as poly(styrene) and poly(methyl methacrylate) is a common commercial practice. DMTA measurements allow a characterisation of the phases present and of the modulus of the resulting material.

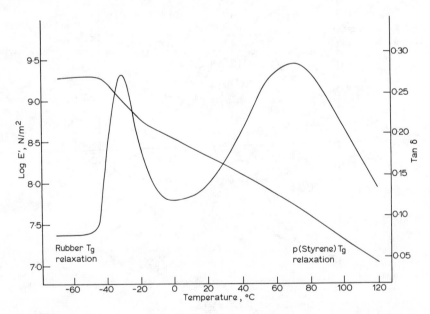

Fig. 6. PL–DMTA scan of block copolymer of butadiene with SBR at 1 Hz and 5 °C min^{-1} heating rate. The two loss peaks clearly show phase separation but with considerable phase boundary mixing.

4.2. Crystalline polymers

Poly(olefins) are commonly produced as copolymers rather than as true homopolymers. The result is frequently a broad melting range. Crystalline polymers exhibit poorly understood relaxation(s) in the region prior to the final melting point, which seems to be related to motions within crystalline phases. These are revealed dramatically in DMTA measurements ($\alpha_{c'}$) but produce little or no enthalpy effects in DSC. Poly(ethylene oxide) is rather easy to study in the high molecular weight form because it is amenable to good clamping. The data in Figs 7 and 8 are for the same poly(ethylene oxide) crystallised under different conditions. The final melting relaxation (α_c) is little affected but the pre-melting relaxation ($\alpha_{c'}$) is changed dramatically.

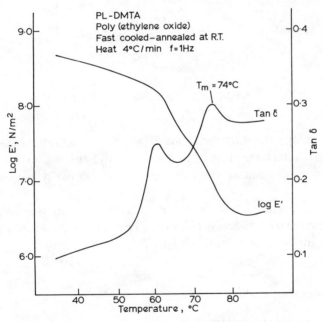

Fig. 7. PL–DMTA scan through the melting range of high molecular weight poly(ethylene oxide), fast cooled from the melt and annealed at room temperature. The pre-melting relaxation is clearly separated from the main melting region.

4.3. Adhesives

Adhesives fall into two categories: (1) rubbery, and (2) thermosetting. The former are used heavily in adhesive tapes and labels, and

R. E. Wetton

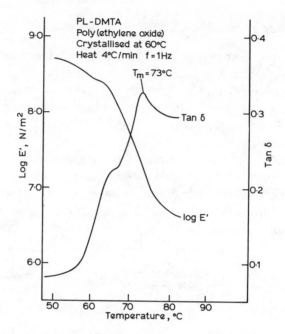

Fig. 8. Sample as in Fig. 7, but crystallised at 60 °C.

the latter in structural applications. The DMTA technique can contribute to characterisation in both areas. In the thermosetting case the changes in modulus and main relaxation are normally studied during curing.

A related area is the curing of adhesives from the liquid state which is extremely difficult to follow. The approach we normally use is to coat an inert substrate and to follow the cure of the resulting composite.

4.4. Engineering Data

Absolute modulus and damping are obtainable from DMTA measurements provided proper attention is paid to clamping and optimising geometry. The results then provide designers with the material parameters they require. Many two-phase materials are strain-dependent, i.e. non-linear visco-elastic. A notable example is carbon-filled rubbers for which data have been obtained in the frequency plane using shear sandwich geometry at different strain levels.

4.5. Frequency Multiplexing

Recent advances in computer control have led to new more-rapid and in some ways more-accurate methods of data acquisition. If during a slow thermal scan, typically at $1\,°C\,min^{-1}$, frequency is continually changed (multiplexed) by an external computer, the data can be sorted and displayed as a series of single measurement frequencies versus temperature. Figures 9 and 10 show this for an epoxy sample through its α process. Here five frequencies have been multiplexed. The peak shift is plotted versus reciprocal temperature in Fig. 11, giving the activation energy, discussed in the theory section as $(-2·303R \times slope) = 383\,kJ\,mol^{-1}$. Rather complete rheological characterisation is given by this technique and, besides the time-saving advantages over isothermal work, two main benefits accrue. Any temperature error is exactly the same for every frequency because the temperature is increasing linearly with time. Furthermore, clamping during heating stays positive as there is a tendency for the sample to expand into the clamps. During isothermal measurements most normal samples tend to relax away from clamping pressure with time. Thus accurate rheological characterisation can also be achieved by this new approach.

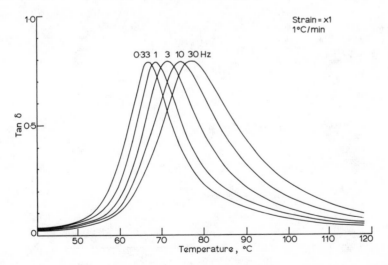

Fig. 9. Frequency multiplexing with the PL–DMTA. The sample was a moderately cross-linked epoxy. This data was acquired at $1\,°C\,min^{-1}$ giving a total time of 80 min.

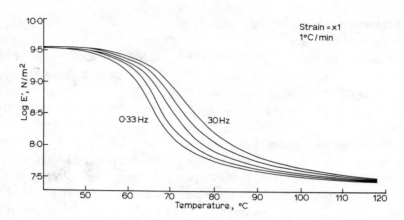

Fig. 10. Frequency multiplexing with the PL–DMTA. The sample was a moderately cross-linked epoxy. This data was acquired at $1\,°C\,min^{-1}$ giving a total time of 80 min.

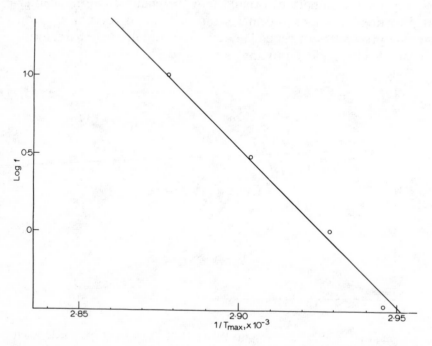

Fig. 11. Loss peak positions plotted as $\log f$ against reciprocal temperature to determine the activation energy of the relaxation process shown in Fig. 10.

5.   FURTHER APPLICATION AREAS

The main application areas in polymer science are as follows:

Engineering data: composites, rubbers
Phase morphology/composites
Adhesive performance
Curing characteristics: epoxies, silicones
Physical ageing
Structure/properties of coatings
Activation energies
Environmental/humidity effects

REFERENCES

1. Zener, C. (1948). *Elasticity and Anelasticity of Metals*, Chicago, Chicago University Press.
2. Alexandrov, A. P. and Lazurkin, Yu. S. (1940). *Acta Physico-chimica URSS*, **12,** 647.
3. Schmieder, K. and Wolf, K. (1953). *Kolloid-Zeit.*, **134,** 149.
4. Cowie, I. *Polymer,* in press.
5. Nolle, A. W. (1950). *J. Polym. Sci.,* **5,** 1.
6. Ferry, J. D. (1961). *Viscoelastic Proproperties of Polymers* New York, John Wiley & Sons Inc.
7. McCrum, N. G., Read, B. E. and Williams, G. (1967). *Anelastic and Dielectric Effects in Polymeric Solids*, New York, John Wiley & Sons Inc.

Polymer Testing **4** (1984) 131–138

Recent Developments in Polymer Characterisation by Dynamic Mechanical Analysis

P. S. Gill, J. D. Lear and J. N. Leckenby*

Du Pont (UK) Ltd, Wedgwood Way, Stevenage, Hertfordshire SG1 4QN, UK

SUMMARY

Examples are given to illustrate the improvements in the Du Pont dynamic mechanical analyser used in conjunction with the Du Pont 1090 thermal analysis system. New techniques and capabilities are discussed and details of the new features are given.

1. INTRODUCTION

Recent developments have provided significant improvements in the capability and ease of operation of the 982 Du Pont dynamic mechanical analyser (DMA) used in conjunction with the Du Pont 1090 thermal analyser system. These improvements in terms of mechanical design, temperature control and data manipulation result in greater accuracy of the measured viscoelastic variables, more precise temperature measurement and control, faster sample throughput and ease of sample and instrument set-up. The ability to handle a broader range of sample geometries has also extended the versatility of the DMA in being able to characterise both very stiff composite samples and very soft elastomer materials.

This paper will provide examples to illustrate these new techniques

* Author to whom correspondence should be addressed.

and capabilities and also provide details of the specifics of these added features.

2. EXPERIMENTAL

Data were generated using the Du Pont 982 DMA in conjunction with the 1090 thermal analyser for control and data analysis. Advanced DMA software was used with automatic system calibration and correction factor calculations.

The Du Pont 982 dynamic mechanical analyser operates on the mechanical principle of forced resonant vibratory motion with fixed selected amplitude. Two parallel balanced sample support arms, free to oscillate around low-hysteresis flexure pivots, are the heart of this system.

The sample is rigidly clamped in place between the two arms, making the arms and sample part of a combined resonant system. The position of one arm and pivot is fixed, while the other arm and pivot are movable by means of a precision mechanical slide to accommodate a wide variety of sample lengths. Once the desired arm spacing is established for the sample, the arm and pivot are locked in place to maintain the two arms in parallel. The system is displaced and set into oscillation by a driver at an amplitude selected by the operator. As the system oscillates, the sample is subjected to a type of deformation similar to that shown in Fig. 1. The amplitude is measured by a linear variable differential transformer (LVDT), the core of which is attached to the fixed-pivot arm. The output of the LVDT contains both the frequency and the amplitude information needed to calculate the viscoelastic properties measured by the 982 DMA.

Normally a system so displaced would oscillate at the system resonant frequency, with constantly decreasing amplitude due to loss of energy (damping) within the sample. The electronics of the 982 are designed to compensate for this loss of energy in the samples. The amplitude signal from the LVDT is fed into a circuit which in turn provides an output signal to the electromechanical driver. This supplies additional energy to the driven arm forcing the system to oscillate continuously at constant amplitude.

The frequency of oscillation is directly related to the modulus of

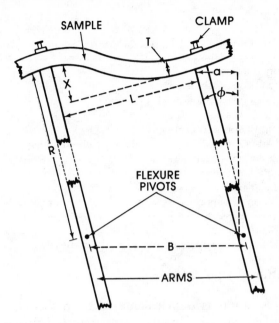

Fig. 1. 982 DMA sample deformation.

the sample under investigation, while the energy needed to maintain constant amplitude oscillation is a measure of damping within the sample.

Quantitative frequency and damping signals are transmitted to the 1090 thermal analyser where they are digitised and stored in disk memory along with sample temperature and time signals. Quantitative calculation of the viscoelastic parameters such as tensile modulus, shear modulus, loss modulus and tan δ is performed using the data analysis software available with the 1090 thermal analyser. Calculated results are subsequently plotted or tabulated on the printer/plotter output.

3. RESULTS AND DISCUSSION

A summary of the various kinds of end-use applications for dynamic mechanical analysis techniques is shown in Table 1. Examples illus-

TABLE 1
End-Use Applications for DMA Techniques

1. Curing of thermosets
2. Polymer blend compatability
3. Correlation of impact stability with damping
4. Observation of plasticiser effects
5. Measurement of subtle transitions
6. Sound and vibration dissipation correlation with damping
7. Engineering design—mechanical data
8. Characterisation of stiff composites and metals
9. Analysis of soft elastomers
10. Characterisation of supported systems (coatings, prepregs, adhesives)

trating some of these applications are given here:

3.1. Engineering design data

Figure 2 shows modulus data calculated from the 982 DMA compared to that obtained by independent determinations for a broad

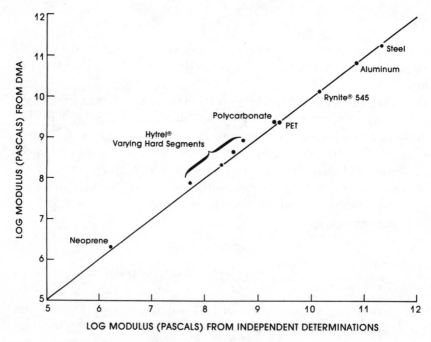

Fig. 2. Comparative modulus.

spectrum of material moduli. A six decade range in modulus was covered from steel at 186 GPa to neoprene at 2·06 MPa. Excellent correlation was shown between the two measurement techniques. Precision data from repeated experiments showed better than ±5% coefficient of variation for modulus values and ±0·005 for tan δ precision for the same series of materials.

3.2. Detection of subtle transitions

Figure 3 shows the DMA output for low-density branched and high-density linear polyethylene. The three transitions shown—alpha, beta and gamma at 80, −3 and −115 °C—are readily detected by the DMA technique but are very difficult to observe by traditional thermal analysis techniques. Each of these transitions corresponds to specific molecular motions which have significance in terms of structure/property relationships:

1. Alpha-transition is associated with crystalline relaxations occurring below the melting point of polyethylene.

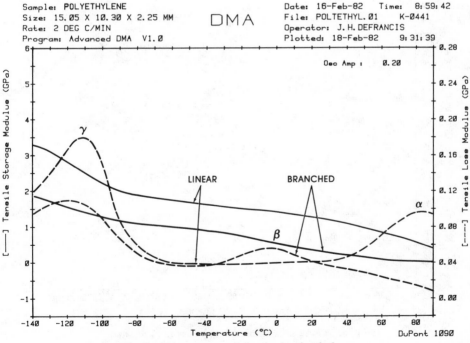

Fig. 3. Linear versus branched polyethylene.

2. Beta-transition is due to motion of the amorphous region side-chains or branches from the main polymer backbone. The intensity of the beta-transition varies with the degree of branching.
3. Gamma-transition is due to crankshaft rotation of short methylene main chain segments and can influence low temperature impact stability of polyethylene.

Lower modulus values are also apparent for the low-density polyethylene compared to the high-density material. Subtle distinctions in the temperatures for modulus changes can also be observed.

3.3. Characterisation of stiff composites

Figure 4 shows the 982 DMA modulus determinations for a series of composite laminates with varying ply orientations and compares them with independent bending modulus calculations made by other techniques. The designation 0/45/90 refers to the orientation of the

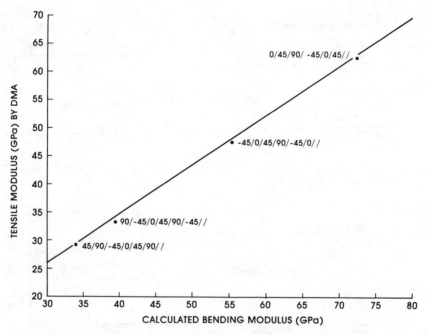

Fig. 4.   Composite laminate—ply orientation effect.

fibres in the ply with respect to the longitudinal direction of the sample. Each data point represents a different stacking arrangement of the oriented fibre plys. The results follow the predicted trend: as the zero-degree ply moves toward the outer surface of the laminate, a progressive increase in modulus is observed. The numbers represent the angle of orientation of the ply relative to the bending axis and each set is a repeat unit as designated by //.

3.4. Thermoset curing behaviour

The viscoelastic changes which take place as a result of cross-linking reactions in a thermosetting polymer can be followed by DMA with respect to both time and temperature. The glass transition temperature, the onset of gelation and vitrification, as well as the modulus and damping of the final cured material, can all be measured by the DMA technique.

An example is shown in Fig. 5. In this experiment, a partially cured graphite-reinforced epoxy prepreg was held horizontally in the instrument and subjected to a simulated cure cycle using the linked method capability of the 1090 programmer. The material was slowly heated to an isothermal pre-cure temperature of 107 °C and held for a predetermined time before being subjected to a final post-cure heat

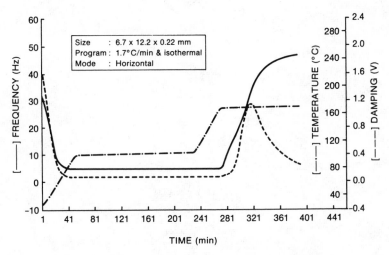

Fig. 5.   Graphite–epoxy prepreg (satin weave) cure cycle.

treatment at 170 °C to achieve optimum cross-linking. The viscoelastic changes occurring during the treatment are interpreted from both the elastic response (resonance frequency) and the damping response. The initial glass transition (T_g), where the uncured resin first softened, is observed at 25 °C as both a damping peak and a frequency decline. This temperature gives an indication of the extent of pre-cure. During the isothermal step, damping and frequency show an increase due to progression of cross-linking reactions. During final cure, a damping peak and a frequency plateau are observed, indicating vitrification as cure completion in the resin. Such information from DMA experiments can be used for optimisation of cure cycles, as well as for investigation of variations in raw material performance.

Polymer Testing **4** (1984) 139

Microstructure of Processed Thermoplastics

M. J. Bevis, J. Bowman and D. Vesely

Department of Non-Metallic Materials, Brunel University, Uxbridge, Middlesex, UK

SUMMARY

The formulation of a plastics compound and the processing conditions used to convert the compound into an artefact together determine its microstructure and physical properties. A wide range of techniques is used to characterise the microstructure of a plastic. Several of these were described with reference to the plastics used for the manufacture of thermoplastics pipeline systems, and included light and electron microscopy, x-ray diffraction and x-ray microanalysis, as applied for the characterisation of polyethylene, polybutylene, polypropylene, ABS and PVC. Reference was made to macroscopic as well as microscopic inhomogeneities in the structure of proposed thermoplastics containing low concentrations of polymer compound additives. One of the examples dealt with in detail related to the structure of unplasticised PVC, and incorporated macroscopic methylene chloride etching and high-resolution transmission electron microscopy studies.

Polymer Testing **4** (1984) 141

Characterisation of Heavily Filled Polymer Compounds

J. Ess, P. R. Hornsby, M. J. Bevis and S. Lin

Department of Non-Metallic Materials, Brunel University, Uxbridge, Middlesex, UK

SUMMARY

It is difficult to measure quantitatively the dispersion of filler particles in heavily filled thermoplastics. Many techniques have been assessed for this purpose and the findings were presented. The presentation was based on an assessment of the dispersion of calcium carbonate in a polypropylene co-polymer, which prior to examination had been extruded into sheet using a co-rotating twin-screw extruder. Details were given of the techniques for specimen preparation, examination and the interpretation of images.

Polymer Testing 0142-9418/84/$03·00 © Elsevier Applied Science Publishers Ltd, England, 1984. Printed in Northern Ireland

Polymer Testing **4** (1984) 143–164

Birefringence Techniques for the Assessment of Orientation

B. E. Read, J. C. Duncan and D. E. Meyer

National Physical Laboratory, Teddington, Middlesex TW11 0LW, UK

SUMMARY

Methods are surveyed for the quantitative specification of molecular orientation in solid polymers based on measurements of birefringence (refractive index anisotropy). Reliable estimates are required of the intrinsic birefringence of the fully oriented material and account should be taken of possible contributions to the birefringence from local distortional deformations and other effects not associated with orientation. The in-plane birefringence of sheets or films is usually determined from measurements of the relative retardation between the resolved components of a polarised light beam transmitted normal to the sample plane. Far-infra-red and microwave techniques can eliminate scattering and yield birefringence values on materials opaque to visible radiation. Several methods are reviewed for measuring the through-thickness birefringence of biaxially oriented sheets or films. From birefringence measurements over a temperature range, using a light-scattering retardation technique, estimates have been made of the rubber–elastic orientational stresses and internal distortional stresses frozen into biaxially oriented sheets of cross-linked PMMA.

1. INTRODUCTION

Many processing methods for polymeric articles yield materials in which the molecular chain axes are preferentially aligned in certain

143

directions. In the case of uniaxially drawn fibres and biaxially stretched sheets or films, this molecular orientation is deliberately introduced to enhance properties such as stiffness, strength and resistance to crazing. On the other hand, regions of high orientation and associated internal stress can give rise to local structural weakness in certain directions and to dimensional instability.

Owing to the structural complexities of most polymeric materials an adequate quantitative characterisation of the state of orientation is often difficult and may require the application of several complementary techniques. These include[1,2] birefringence, x-ray pole figure analysis, polarised spectroscopy and sonic velocity methods. Novel methods for assessing the local orientation in plastic components have also recently been proposed based on indentation fracture tests[3] and thermal probe analysis,[4] respectively.

Of the above methods, the birefringence technique is particularly well known and can yield well-defined measures of orientation using readily available equipment. In this survey the principles, scope and limitations of this method will therefore be discussed giving some emphasis to recent developments at the National Physical Laboratories (NPL) and elsewhere.

2. SPECIFICATION AND ORIGIN OF BIREFRINGENCE

2.1. General relationships

Birefringence in solid polymers[5] most frequently arises from the orientation of structural units produced by the application of a stress to the material in the rubber-like or partially crystalline state. The structural unit may correspond to a link in an amorphous chain molecule or to a crystallite axis, and its orientation may be specified with respect to reference axes 1, 2 and 3 in the sample (Fig. 1) where axis 1 is taken to denote the thickness direction for plastic sheets or films. Assuming the unit to have cylindrical symmetry, θ is used to denote the angle between the axis of the unit and the reference direction 3, and ϕ is the azimuthal angle which the projection of the axis on the 1–2 plane makes with direction 1. The polarisabilities $a_{\parallel}$ and $a_{\perp}$ (defined as the induced dipole moments per unit electric field) represent the ease with which the charge distribution in the unit is

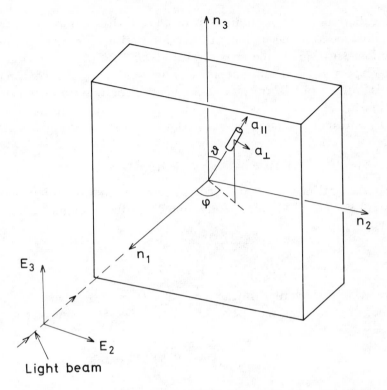

Fig. 1. Illustration of the method for specifying the orientation of a structural unit and refractive indices.

displaced by an electric field polarised parallel and perpendicular to the unit axis, respectively.

Within an oriented polymer sample the structural units will be preferentially aligned over certain ranges of the angles θ and ϕ, and as a consequence the net sample polarisability will depend on the direction of the applied electric field. Owing to the relationship between polarisability and refractive index, it follows that the refractive indices n_1, n_2 and n_3 for radiation polarised in the respective directions may not, in general, be equal. If, for example, a polarised light beam is transmitted through the sample along direction 1 (Fig. 1) then, assuming that $n_3 > n_2$, its velocity within the sample will be smaller if it is vertically polarised (electric field E_3) than if horizontally polarised (electric field E_2).

So-called 'uniaxial orientation' is produced when a polymer strip or rod is extended along its length direction and, if this corresponds to direction 3 in Fig. 1, the structural units will be randomly oriented with respect to ϕ. From the tensor summation of the component polarisabilities of all units[6] and application of the Lorentz–Lorenz relationship between refractive index and polarisability it follows that $n_1 = n_2$ and that a single non-zero component of birefringence Δn is given by

$$\Delta n = n_3 - n_2 = n_3 - n_1 = \Delta n_0 f \tag{1}$$

where the intrinsic birefringence Δn_0 is the limiting birefringence when all structural units are fully aligned in the stretching direction and is given by

$$\Delta n_0 = \frac{2\pi}{9} \frac{(n^2 + 2)^2}{n} N(a_\parallel - a_\perp) \tag{2}$$

where n is the average refractive index $(n_3 + 2n_2)/3$, and N is the number of structural units per unit volume. The Hermans orientation function f is given by

$$f = \frac{3 \overline{\cos^2 \theta} - 1}{2} \tag{3}$$

where $\overline{\cos^2 \theta}$ is the value of $\cos^2 \theta$ averaged over the orientation of all units. We may note that the value of f varies from zero for the unoriented sample (completely random orientation of units) to unity for the fully oriented material.

If a sheet of polymeric material is extended by unequal amounts along the two in-plane orthogonal directions 2 and 3 (Fig. 1), then n_1, n_2 and n_3 will differ from each other and the material will exhibit biaxial orientation. From the tensor summation of polarisability components we now obtain

$$n_3 - n_2 = \Delta n_0(f + g/2) \tag{4a}$$

$$n_3 - n_1 = \Delta n_0(f - g/2) \tag{4b}$$

$$n_1 - n_2 = \Delta n_0 g \tag{4c}$$

where $g = \overline{\sin^2 \theta \cos 2\phi}$ which is the value of $\sin^2 \theta \cos 2\phi$ averaged over all units. In general, we may thus write eqn. (4) in the form

$$n_i - n_j = \Delta n_0 F_{ij} \qquad i, j = 1, 2 \text{ or } 3 \tag{5}$$

in which F_{ij} is an orientation parameter involving averages of functions of θ and ϕ.

According to eqns. (1) and (4) the orientation functions f and g could be obtained from measurements of birefringence if Δn_0 is known. Calculations of intrinsic birefringence may, however, be limited by uncertainties in the conformations of structural units, internal field effects and the values to be assigned to bond polarisability components.[5,7] Experimental determinations of Δn_0 from extrapolations of birefringence data to high extensions are possible and may be aided by complementary measurements of other properties such as sonic velocity.[1,8] It should also be emphasised that birefringence measurements cannot yield the complete orientation distribution of structural units but only the mean squared functions represented by f and g.

2.2. Rubber-like amorphous polymers

In considering the stress-induced birefringence of amorphous polymers a distinction must be made between rubbery and glassy materials owing to basic differences between the deformation mechanisms in the respective states. Theories of birefringence in cross-linked rubbers stem from the model of Kuhn and Grün[9] in which each polymer chain between cross-link points in the network comprises many freely jointed statistical links. Each link represents a structural unit and is free to assume any orientation independent of neighbouring links. Polarisability components $a_{\parallel}$ and $a_{\perp}$ are assigned to each link for light polarised parallel and perpendicular to the link axis, respectively. Upon extension, the components of each network chain end-to-end vector deform affinely (i.e. in the same ratio as the bulk sample dimensions) and the chain links orient toward the directions of extension. For a general homogeneous strain characterised by extension ratios α_i and α_j in principal directions i and j the theory yields[10]

$$n_i - n_j = \frac{2\pi}{45} \frac{(n^2+2)^2}{n} N_c(a_{\parallel} - a_{\perp})(\alpha_i^2 - \alpha_j^2) \tag{6}$$

where N_c is the number of network chains per unit volume.

On the basis of the above model, the kinetic theory of rubber elasticity[10] gives the following well-known equation for the difference between any two entropy-elastic principal stress components σ_i

and σ_j

$$\sigma_i - \sigma_j = N_c k T (\alpha_i^2 - \alpha_j^2) \qquad (7)$$

in which k is Boltzmann's constant and T is the absolute temperature.

Equation (6) may now be expressed in the form of eqn. (5)

$$n_i - n_j = \Delta n_0 F_{ij}$$

$$= \frac{2\pi}{9} \frac{(n^2+2)^2}{n} N_l (a_\parallel - a_\perp) F_{ij} \qquad (8)$$

where N_l is the number of statistical links per unit volume and hence the orientation function becomes

$$F_{ij} = \frac{1}{5} \frac{N_c}{N_l} (\alpha_i^2 - \alpha_j^2) = \frac{\sigma_i - \sigma_j}{5 N_l k T} \qquad (9)$$

According to eqn. (9) the orientation of statistical links, as characterised by F_{ij}, is predicted to be proportional to the *stress* difference $\sigma_i - \sigma_j$. The strain function $\alpha_i^2 - \alpha_j^2$, and consequently the subsequent retraction or shrinkage on removal of external loads, is not simply related to F_{ij} but also depends on the degree of cross-linking through N_c. For uncross-linked materials an effective cross-linking is provided by chain entanglements if, subsequent to deformation, insufficient time is allowed for the structure to relax to equilibrium. Some caution should therefore be exercised in interpreting the results of standard shrinkage tests[11] on amorphous thermoplastics in terms of the state of orientation.

The ratio of birefringence to stress difference for rubbers defines an 'orientational' stress–optical coefficient C_{or} which, from eqns. (6) and (7), is given by

$$C_{or} = \frac{n_i - n_j}{\sigma_i - \sigma_j} = \frac{2\pi}{45 k T} \frac{(n^2+2)^2}{n} (a_\parallel - a_\perp) \qquad (10)$$

The orientational stress–optical coefficient is thus predicted to be independent of the degree of cross-linking, as found experimentally, and proportional to the link anisotropy $a_\parallel - a_\perp$.

The above theory is, of course, based on an idealised model and it is not usually possible to accurately identify the statistical link with any real conformational sequence of chain bonds. Modern theories of stress and birefringence of rubber-like polymers are based on more realistic models including the fixed valence angles and rotational isomeric states of main-chain bonds.[12,13] These theories show that eqn. (10) is still valid if $a_\parallel - a_\perp$ is now identified with an effective link anisotropy obtained from the tensor additivity of component bond polarisabilities averaged over the unperturbed chain. The theories have been successfully applied to rubbers swollen with isotropic solvent molecules and can be adapted to undiluted systems.[7] According to these theories a varation of $a_\parallel - a_\perp$, and hence C_{or}, with temperature can arise from equilibrium energy differences between different rotational isomeric states.

2.3. Amorphous polymers in the glassy state and the glass–rubber transition region

Elastic deformation mechanisms in polymeric glasses are not considered to involve principally the orientation of chain links but rather the local twisting around main-chain bonds against rotational energy barriers and changes in intermolecular spacing against the forces of attraction or repulsion. These local deformations give rise to energy–elastic 'distortional' birefringence Δn_d and to distortional stresses $\Delta \sigma_d$. Although the distortional birefringence is not well understood on a molecular basis, anisotropic local field effects appear to yield a significant birefringence contribution[14] and the local tilting of side-groups towards the stretching direction can produce a net distortional birefringence of either positive or negative sign.[15,16] Available evidence suggests that the equilibrium distortional stress–optical coefficient $C_d = \Delta n_d / \Delta \sigma_d$ may be little affected by variations of temperature.[14]

In the temperature range associated with the transition from rubbery to glassy behaviour, the mechanism of deformation will partly involve segmental orientation characteristic of a rubber and partly the local glass-like distortions. The relative contributions from the two mechanisms will depend on the time-scale of loading which is now comparable with the relaxation times for molecular rearrangements associated with the glass–rubber transition. Consequently, the measured stress–optical coefficient at low strains will depend on the loading rate in a force–elongation test or on the duration of loading in a creep or stress–relaxation test and will vary from the distortional value C_d at short times to the equilibrium orientational value C_{or} at long times (Fig. 2).

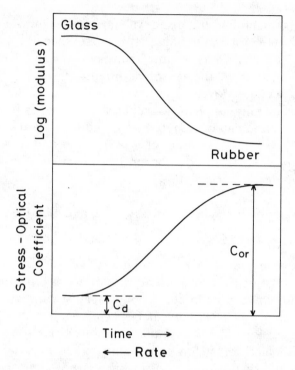

Fig. 2. Typical variation of modulus and stress–optical coefficient with time or rate of loading for an amorphous polymer in the glass–rubber transition region.

From birefringence measurements during creep, stress relaxation or dynamic testing it is possible to resolve the time- and frequency-dependent stresses (or moduli) into their orientational and distortional components.[16,17] We have recently proposed[18] a related method for analysing the non-equilibrium stresses 'frozen' into certain glassy polymers as a result of stretching in the rubbery state at high temperatures and cooling in the deformed state to temperatures below the glass-transition temperature T_g. In this method the birefringence $\Delta n(T)$ measured at temperature T below T_g is expressed in the form

$$\Delta n(T) = C_{or}(T)\,\Delta\sigma_{or} + C_d(T)\,\Delta\sigma_d \qquad (11)$$

where $\Delta\sigma_{or}$ and $\Delta\sigma_d$ are components of the frozen stress differences $\sigma_i - \sigma_j$, and $C_{or}(T)$ and $C_d(T)$ are the limiting stress–optical coefficients (Fig. 2) written as a function of temperature. If $C_{or}(T)$ has a significant temperature dependence at $T < T_g$ and the temperature dependence of $C_d(T)$ is assumed negligible, then $\Delta\sigma_{or}$ and $\Delta\sigma_d$ can be estimated from the respective slope and intercept of an experimental plot of $\Delta n(T)$ against $C_{or}(T)$. This analysis is applied in Section 4 to birefringence data on stretched cross-linked polymethyl methacrylate (PMMA).

2.4. Partially crystalline polymers and blends

Birefringence measurements can provide some 'average' assessment of orientation in multicomponent systems but cannot distinguish between the relative levels of orientation in the different phases unless complementary measurement techniques are also employed. In the case of partially crystalline polymers, for example, an amorphous phase orientation function can be obtained from birefringence data if the crystallite orientation is known from another technique such as x-ray analysis. This approach assumes that birefringences arising from the different phases are additive and should take account of form birefringence and deformation effects other than orientation.[5]

3. MEASUREMENT OF THE IN-PLANE BIREFRINGENCE $n_3 - n_2$

3.1. Visible wavelengths

Figure 3 illustrates a typical optical system for birefringence determinations using visible radiation (wavelength range $0 \cdot 4 - 0 \cdot 8 \ \mu$m). For measurements of $n_3 - n_2$ on flat sheets or strips, the sample is located between crossed polarisers (polaroid sheets or prisms) in a parallel beam of light which is incident normal to a face of the specimen. The sample is inclined with principal axes 2 and 3 at 45° to the polarisation axes of the two polarisers. Light incident on the sample is thus plane polarised with (transverse) electric field vector at 45° to axes 2 and 3 and within the material the beam may be resolved into two equal amplitude components polarised with electric vectors along 2 and 3, respectively. Owing to the difference in magnitude between n_2 and n_3 the component waves travel with different velocities or wavelengths

B. E. Read, J. C. Duncan, D. E. Meyer

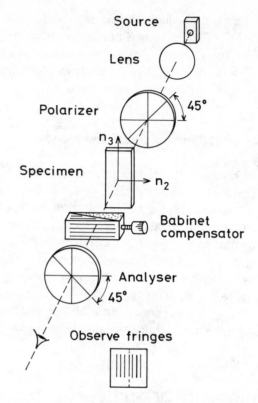

Fig. 3. Optical system for measuring phase retardations at visible wavelengths.

and emerge from the sample with a relative phase shift or retardation. If a monochromatic beam is employed of wavelength λ_0 in vacuum, and d is the sample thickness, the relative retardation R_{32} is related to the birefringence by[2]

$$n_3 - n_2 = R_{32}\lambda_0/d \tag{12}$$

If I_0 is the intensity of polarised light incident on the sample then, in the absence of the compensator, the intensity I_t transmitted through the second polariser (or analyser) is given by[19]

$$I_t = I_0 \sin^2 (\pi R_{32}) = \frac{I_0}{2}(1 - \cos 2\pi R_{32}) \tag{13}$$

From eqn (13) R_{32} values less than one wavelength could be determined from measurements of transmitted intensity and $n_3 - n_2$ subsequently obtained using eqn. (12). This method is useful if the birefringence is changing rapidly with time.

For static retardation measurements, a compensator is conveniently employed in the optical beam between the sample and analyser. Compensators may comprise birefringent plates which are either rotated or tilted within the light beam. Figure 3 illustrates the well-known Babinet compensator which comprises two quartz wedges cut with their optic axes perpendicular to each other and to the light beam direction. When these axes are aligned at ±45° to the polariser and analyser axes, a series of equally spaced fringes is observed in the transmitted light using monochromatic radiation. The fringes arise from the variation of net retardation across the beam and the compensator is calibrated with monochromatic light by measuring the lateral movement of one wedge per fringe displacement. With a white-light source, coloured fringes are observed either side of a central black fringe (in the absence of a sample) owing to the variation of retardation with wavelength. The black zero-order fringe will still be evident but shifted laterally when a sample is introduced into the beam if the retardations of the sample and compensator have a similar wavelength dependence. The zero-order fringe shift then provides a measure of the sample retardation. Data obtained by this method for three oriented thermoplastics are included in Table 1 and commented on in Section 3.2.

Figure 4 illustrates an experimental set-up recently employed at the NPL for determining the local birefringence values and low-angle light scattering characteristics of material in the vicinity of heat-seals in tubular extruded low-density polyethylene (LDPE) films. A sample of film containing the heat-seal was mounted in a goniometer and located in the path of a polarised He–Ne laser beam together with a compensator and polaroid sheet analyser. In Fig. 4 the heat-seal lies perpendicular to the plane of the diagram and to the extrusion direction (machine direction) of the film. The heat-seal had a width of about 1 mm and thickness of 0·33 mm compared with a film thickness of 0·065 mm away from the seal. The laser beam was polarised with electric vector at 45° to the sample machine direction and, after passing through a lens, slowly converged to a diameter of about 0·2 mm at the sample. This enabled the beam to pass directly through the seal. The compensator was of the Babinet–Soleil type, comprising two birefringent wedges with parallel optic axes and a birefringent

TABLE 1

Birefringence Values $n_3 - n_2$ Measured at Various Wavelengths

	Wavelength			
	Visible	*Far-infra-red*		*Microwave*
Polymer	*0·546 μm*	*0·5 mm*	*1 mm*	*8·6 mm*
Polyethylene				
Low-density sheet; draw ratio 1·4; thickness 0·333 mm	0·016	0·016	0·016	0·015
Poly(ethylene-terephthalate)				
Biaxial Melinex film; thickness 0·058 mm	0·034	0·006	0·012	0·035
Polystyrene				
Drawn at 118°C; Frozen stress $=5·5\times10^5\,N\,m^{-2}$; thickness 3·24 mm	−0·0033			−0·0028
PTFE				
Drawn at 253°C; draw ratio 2·4; thickness 4·25 mm				0·0487

plate with perpendicular optic axis. This produced a uniform retardation across the light beam which was varied, to compensate for the sample retardation, by adjusting the lateral position of one wedge until extinction of the transmitted beam was observed on the screen. From the known retardation introduced by the calibrated compensator the sample retardation and birefringence were calculated (eqn. (12)).

The low-angle light scattering intensity distribution produced by the heat-seal material was observed on the screen after removing the compensator (Fig. 4(b)). So-called 'H$_V$ light scattering patterns' were obtained, where H and V indicate a horizontal polarisation axis of the analyser and vertically polarised incident light respectively. These

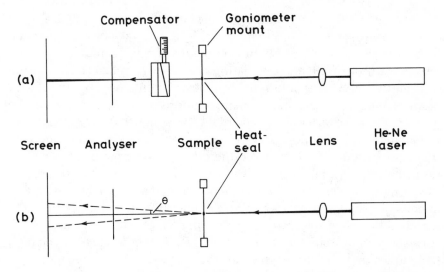

Fig. 4. Measurement of (a) birefringence and (b) light-scattering patterns for material within and close to a plastic heat-seal.

patterns are related to the morphological structure of the LDPE material and indicate in particular whether spherulites are present.[5] The birefringence and light-scattering investigations have shown that the material within the heat-seal has retained a partially oriented non-spherulitic structure similar to that of the original film away from the seal. The apparently small changes in orientation and morphological structure occurring during heat-sealing are consistent with the satisfactory seal strengths observed in LDPE.

3.2. Far-infra-red and microwave birefringence

Birefringence measurements at visible wavelengths are usually confined to relatively thin films in the case of partially crystalline or blended polymers. Above a certain thickness, the amount of light scattered by such materials is usually sufficient to render them opaque to visible radiation. An extension of birefringence measurements to longer wavelengths[2,20] should make it possible to overcome this limitation owing to the appreciable decrease in scattering with increasing wavelength.

At wavelengths corresponding to the far-infra-red ($\lambda_0 \approx 0 \cdot 1$–2 mm) and microwave ($\lambda_0 \approx 3$–300 mm) regions of the electromagnetic spectrum, the scattering of light is generally insignificant and both the resonance absorption and dissipation of energy by relaxation are often sufficiently small to enable reliable measurements of refractive index or birefringence to be made. Contributions to $a_{\parallel}$ and $a_{\perp}$ from the displacements of atomic nuclei might be somewhat larger at these wavelengths than in the visible region but these contributions may be small compared with those from the electronic charge displacements, resulting in similar values of birefringence in the respective wavelength ranges. However, the wavelengths of far-infra-red and microwave radiation are between 10^3 and 10^5 times longer than the wavelengths of visible radiation and net phase retardations will be correspondingly smaller (eqn. (12)). Thus the longer wavelength techniques may provide less accurate measures of birefringence than the visible method.

At far-infra-red wavelengths refractive indices can be determined, together with the corresponding absorption coefficients, from the Fourier transformation of the outputs of a Michelson interferometer.[21,22] If a polariser is placed in the combined interferometer beams prior to reaching the detector, then $n_3 - n_2$ can be obtained[20] from the two interferograms recorded with the sample inserted in one beam and the polariser axis successively aligned parallel to the principal sample directions 3 and 2. Table 1 includes some comparative data at wavelengths of $0 \cdot 5$ mm and 1 mm obtained recently by dispersive Fourier transform spectrometry using a modular Michelson interferometer developed at the NPL.[23]

In the microwave region, values of n_3, n_2 and hence $n_3 - n_2$ may be determined using an open resonator technique.[24] Table 1 includes $n_3 - n_2$ data obtained at the NPL by this method with $\lambda_0 \approx 8 \cdot 6$ mm.

In Table 1 the $n_3 - n_2$ values for polyethylene and polyethylene terephthalate (Melinex) are accurate to within about $0 \cdot 001$ whilst for the thicker polystyrene and polytetrafluoroethylene (PTFE) samples a somewhat higher accuracy ($0 \cdot 0005$) is estimated. Although the visible birefringence values have a potentially higher level of accuracy, sample non-uniformities gave rise to variations in the measured values with position and the quoted data were averaged over the sample areas. It will be noted from Table 1 that for drawn polyethylene and polystyrene good agreement is obtained between the birefringence values at different wavelengths. In the case of Melinex, the visible and microwave values agree well but the far-infra-red values are considerably lower and are frequency-dependent. This result probably arises from a strong absorption process and corresponding dispersion of refractive indices in the far-infra-red region. Birefringence measurements were not possible at visible wavelengths for the thick opaque PTFE sample but this material was suited to the microwave resonance technique owing to its low dielectric loss. PTFE is also suitable for investigation by the far-infra-red technique although measurements on this oriented sample have not been made.

4. MEASUREMENT AND ANALYSIS OF BIREFRINGENCE COMPONENTS $n_3 - n_1$ AND $n_2 - n_1$

Values of $n_3 - n_1$ and $n_2 - n_1$ for biaxially oriented sheets may be obtained from measurements of R_{31} and R_{21} using

$$n_i - n_1 = R_{i1}\lambda_0/d \qquad i = 2 \text{ or } 3 \qquad (14)$$

R_{i1} is thus the relative retardation between the resolved components of a polarised beam having electric field vectors along directions i and 1, respectively. It follows that for measurement of R_{i1} the beam cannot be transmitted in a direction normal to the 2–3 plane of the sheet.

Relatively thin sheets or films have been conveniently investigated[2] by rotating the specimen in the optical beam (Fig. 3) about either of the axes 2 or 3. Retardation measurements as a function of the angle of tilt then allow the determinations of $n_3 - n_1$ (rotation about axis 2) and of $n_2 - n_1$ (rotation about axis 3).

Solvent-cast films of thickness about 3 μm have been studied by a novel technique in which a narrow laser beam is coupled into and out of the film by means of a prism.[25] At discrete angles of incidence, waves are excited within the film which acts as a waveguide for allowed optical modes. Measurement of these angles for modes with the transverse component of the electric vector in and perpendicular to the plane of the film allows the determination of n_3 (or n_2) and n_1 and hence the differences $n_3 - n_1$ or $n_2 - n_1$, respectively.

Measurements of the individual components n_1, n_2 and n_3 have also been made with a modified Abbé refractometer on various anisotropic polymer films.[26]

In the case of sheets with thickness greater than a few millimetres $n_3 - n_1$ or $n_2 - n_1$ may be determined with an incident light beam directed perpendicular to the 3–1 or 2–1 cross-section of the sheet. For this purpose, strips may be machined from the sheet having faces parallel to the sheet cross-section and, after polishing, located between crossed polarisers such that the strip faces are normal to the incident light beam. The transmitted beam in Fig. 3 is then resolved into equal amplitude components polarised with the electric vector along directions 3 (or 2) and 1. Local measurements of retardation with this arrangement can reveal any variations of $n_3 - n_1$ or $n_2 - n_1$ across the sheet thickness, unlike the methods described above. The technique is, of course, destructive and the machining process can influence residual distortional stress contributions to the birefringence if these are present. This fact has been utilised in separating contributions to the birefringence associated with molecular orientation and residual distortional stress.[27]

Through-thickness variations of $n_3 - n_1$ and $n_2 - n_1$ may also be studied for thick sheets of transparent plastics by a non-destructive technique based on observations of the light scattered by the material from a propagating polarised beam. This method has been employed at the NPL for investigating biaxially stretched sheets of cross-linked PMMA and is illustrated in Fig. 5. The rectangular plastic sheet has a polished edge to minimise surface scattering and is supported with its main faces at 45° to the vertical. A vertically polarised beam from a 20 mW He–Ne laser is first passed through a cylindrical lens to form a sheet of light which spans the thickness of the plastic sheet The beam then enters through the polished edge of the plastic sheet and is transmitted in a direction normal to the sheet cross-section. Within the material the vertically polarised beam is resolved into two equal amplitude components polarised parallel to the plane of the sheet (direction 3 or 2) and parallel to the thickness direction 1. These components becomes progressively out of phase by an amount R_{i1} determined by $n_i - n_1$ and by the distance travelled by the beam in the sheet (eqn. (14)). Variations in retardation R_{i1} give rise to intensity variations in the scattered light which are photographed in a horizontal direction at right angles to the transmitted beam. Since this direction is perpendicular to the polarisation direction of the

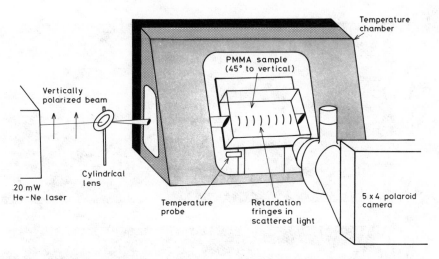

Fig. 5. Schematic illustration of apparatus for recording light-scattering retardation fringes.

incident beam and at 45° to the principal axis 1 it can be shown[28-30] that the scattered intensity I_s is given by

$$I_s \propto 1 + \cos 2\pi R_{i1} \qquad (15)$$

which varies periodically with R_{i1} and hence with distance along the beam within the material. The alternating intensity will give rise to retardation fringe patterns in the scattered light originating from different points within the beam. The distance between neighbouring fringes or intensity maxima is that required to produce one wavelength relative retardation.

Examples of scattered light retardation fringes for cross-linked PMMA samples are shown in Fig. 6. The photograph in Fig. 6(a) was obtained for a cast unoriented sheet which had been rapidly quenched from 140 °C into water at room temperature. The widely spaced curved fringes correspond to low values of $n_3 - n_1$ or $n_2 - n_1$ which vary across the sheet thickness symmetrically with respect to the central plane. This result is consistent with the rapid non-uniform cooling through the glass-transition temperature region since this will produce residual distortional stresses, and possibly some low levels of orientation, which vary in magnitude across the sheet thickness.[27,31] In the case of PMMA the dominant contribution to this birefringence should come from the distortional stresses which are characteristically

B. E. Read, J. C. Duncan, D. E. Meyer

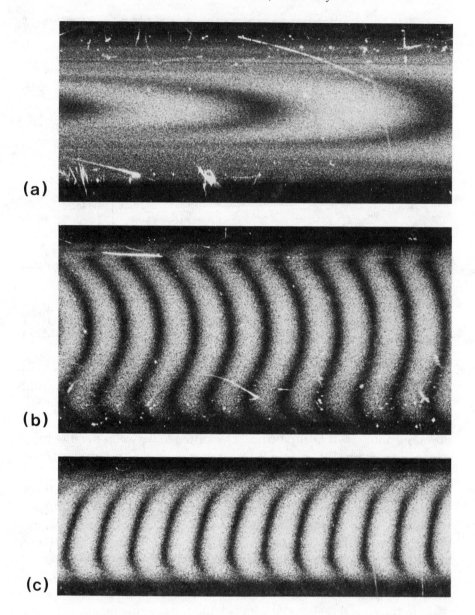

Fig. 6. Light-scattering retardation fringes obtained for (a) quenched cast specimen, (b) hot-stretched biaxially oriented material, and (c) biaxially oriented aircraft
window. Photographs taken part way along the transmitted beams within the
materials with the beam entering from the right.

compressive at the surface and tensile along the central plane. These stresses are physically active in the glass and produce the well-known curvature when the balance of forces is disturbed by the removal of surface layers of material.[32]

The fringes shown in Fig. 6(b) originate from a sheet which had been biaxially stretched by about 65% in the rubbery state and subsequently cooled in the deformed state to below T_g. The closely spaced fringes correspond to larger values of $n_3 - n_1$ and $n_2 - n_1$ arising from the molecular orientation induced in the rubber and frozen into the glassy plastic. The fringe pattern in Fig. 6(c) was obtained for an aircraft window which was fabricated from stretched material by forming it into a slightly curved shape after re-heating to temperatures just above T_g. The closely spaced fringes illustrate that the orientation is retained during the forming process. However, a precise interpretation of the observed fringe patterns requires a consideration of the respective contributions to the birefringence from frozen segmental orientation and internal distortional stress.

In Fig. 7 birefringence data $\Delta n(T)$ obtained for a biaxially stretched sheet of cross-linked PMMA at temperatures between -40 and $+60\,°C$ are plotted against $C_{or}(T)$. The $n_3 - n_1$ and $n_2 - n_1$ values were measured by the light-scattering technique with the transmitted beam along directions 2 and 3, respectively. The $C_{or}(T)$ values were obtained from birefringence measurements using a Babinet compensator on strips of oriented polymer over a temperature range below T_g. The strips were initially oriented by stretching at $160\,°C$ and after attaining equilibrium they were slowly cooled under known constant load to $T < T_g$ and the load was subsequently removed.

The slopes of the two plots in Fig. 7 are very similar and yield, according to eqn. (11), orientational stress differences $(\sigma_3 - \sigma_1)_{or}$ and $(\sigma_2 - \sigma_1)_{or}$ of $2·1\,MN\,m^{-2}$. For the distortional stress differences the two intercepts give $(\sigma_3 - \sigma_1)_d = 7·7\,MN\,m^{-2}$ and $(\sigma_2 - \sigma_1)_d = 3·0\,MN\,m^{-2}$ using $C_d = -6·5 \times 10^{-12}\,N^{-1}\,m^2$. This value of C_d is approximate, having obtained from stress-birefringence studies on the glassy polymer allowing for time-dependent effects associated with the secondary β-relaxation process in PMMA. The scatter of data and long extrapolations in Fig. 7 also render the estimates of distortional stress approximate and it remains to be confirmed whether the differences between $n_3 - n_1$ and $n_2 - n_1$ are predominantly due to distortional stress differences or to small differences in orientation.

B. E. Read, J. C. Duncan, D. E. Meyer

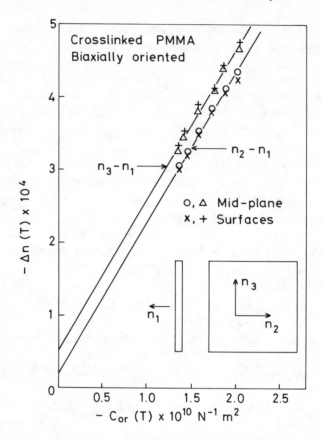

Fig. 7. Plots of birefringences $\Delta n(T)$ against orientational stress–optical coefficient $C_{or}(T)$, each determined at temperatures between -40 and $+60\,°C$, for a biaxially stretched sheet of cross-linked PMMA.

It should be noted that the above analysis yields stress *difference* components rather than absolute stresses. However, we may take σ_1 components to be zero in all cases and from the signs of the slopes and intercepts in Fig. 7 the orientational and distortional components of σ_2 and σ_3 are tensile. It should further be emphasised that, whereas the orientational rubber–elastic stresses exist at temperatures above T_g, they may not be physically active in the glass but characteristic of molecular orientation frozen into the glass by virtue of the very long recovery times at $T < T_g$ for large-scale chain rearrangements. The dominant parts of the distortional stresses may also be inactive in the glass and associated with high-energy chain distortions induced by the stretching and/or cooling and having long recovery times at $T < T_g$. The results in Fig. 7 indicate that these distortional stresses do not vary across the sheet thickness and they may not therefore be of the active instantaneous kind discussed in connection with Fig. 6(a). The origins and measurement of various types of internal stress are discussed by White.[33]

ACKNOWLEDGEMENTS

The authors wish to thank Drs J. R. Birch and R. G. Jones of the Electrical Science Division, National Physical Laboratory, for their birefringence determinations in the far-infra-red and microwave regions.

REFERENCES

1. Ward, I. M. (Ed.) (1975). *Structure and Properties of Oriented Polymers*, London, Applied Science Publishers Ltd.
2. Read, B. E. (1976). *Plastics and Rubber Materials and Applications*, **1**, 123.
3. Kent, R. J., Puttick, K. E. and Rider, J. G. (1981). *Plastics and Rubber Processing and Applications*, **1**, 55, 111.
4. Berrie, M. A., Puttick, K. E., Rider, J. G., Rudman, M. and Whitehead, R. D. (1981). *Plastics and Rubber Processing and Applications*, **1**, 129.
5. Stein, R. S. and Wilkes, G. L. (1975). In: *Structure and Properties of Oriented Polymers*, I. M. Ward (Ed.), London, Applied Science Publishers.
6. Nomura, S., Kawai, H., Kimura, I. and Kagiyama, M. (1967). *J. Polymer Sci., Part A-2*, **5**, 479.
7. Stein, R. S. and Hong, S. D. (1976). *J. Macromol. Sci. – Phys.* **B12**(1), 125.
8. Samuels, R. J. (1965). *J. Polymer Sci., A*, **3**, 1741.
9. Kuhn, W. and Grün, F. (1942). *Kolloid-Z*, **101**, 248.
10. Treloar, L. R. G. (1967). *The Physics of Rubber Elasticity*, 2nd edn, Oxford, Oxford University Press.
11. ISO 2557/1. (1976). BS 2782, Part 9, Method 940 A. (1981). *Preparation of Test Specimens of Amorphous Thermoplastic Moulding Material with a Defined Level of Shrinkage*.
12. Flory, P. J. (1969). *Statistical Mechanics of Chain Molecules*, New York, Interscience.

13. Nagai, K. (1964). *J. Chem. Phys.*, **40**, 2818.
14. Pick, M. and Lovell, R. (1979). *Polymer*, **20**, 1448.
15. Rudd, J. F. and Gurnee, E. F. (1957). *J. Appl. Phys.*, **28**, 1096.
16. Read, B. E. (1970). *Proc. 5th Intern. Congr. Rheology*, **4**, 65.
17. Priss, L. S., Vishnyakov, I. I. and Pavlova, I. P. (1980). *Intern. J. Polymeric Mater.*, **8**, 85.
18. Duncan, J. C. and Read, B. E. (1984) To be published.
19. Mindlin, R. D. (1937). *J. Opt. Soc. Amer.*, **27**, 288.
20. Jacobsson, S. and Hård, S. (1982). *Rev. Sci. Instrum.*, **53**, 1012.
21. Birch, J. R. and Parker, T. J. (1979). In: *Infrared and Millimetre Waves*, Vol. 2, K. J. Button (Ed.), New York, Academic Press.
22. Birch, J. R., Dromey, J. D. and Lesurf, J. (1981). *The Optical Constants of some Common Low Loss Polymers between 4 and 40 cm^{-1}*, NPL Report, DES 69, February.
23. Chantry, G. W., Evans, H. M., Chamberlain, J. and Gebbie, H. A. (1969). *Infrared Phys.*, **9**, 85.
24. Jones, R. G. (1976). *J. Phys. D, Appl. Phys.*, **9**, 819.
25. Prest, W. M. and Luca, D. J. (1979). *J. Appl. Phys.*, **50**, 6067.
26. Samuels, R. J. (1981). *J. Appl. Polymer Sci.*, **26**, 1383.
27. Saffell, J. R. and Windle, A. H. (1980). *J. Appl. Polymer Sci.*, **25**, 1117.
28. Srinath, L. S. and Frocht, M. M. (1963). In: *Symposium on Photoelasticity*, M. M. Frocht (Ed.), Oxford, Pergamon, p. 277.
29. Jessop, H. T. (1951). *Brit. J. Appl. Phys.*, **2**, 249.
30. Gayles, J. N., Lohmann, A. W. and Peticolas, W. L. (1967). *Appl. Phys. Letters*, **11**, 310.
31. Struik, L. C. E. (1978). *Polym. Eng. and Sci.*, **18**, 799.
32. Haworth, B., Hindle, C. S., Sandilands, G. J. and White, J. R. (1982). *Plastics and Rubber Processing and Applications*, **2**, 59.
33. White, J. R. (1984). *Polymer Testing*, this volume, pp. 165–91.

Polymer Testing **4** (1984) 165–191

Origins and Measurement of Internal Stress in Plastics

J. R. White

Department of Metallurgy and Engineering Materials, University of Newcastle upon Tyne, Newcastle upon Tyne NE1 7RU, UK

SUMMARY

The term 'internal stress' is given various meanings in the literature and these are listed and discussed. The origin of deformation-induced internal stress is considered, and the development of residual (moulding) stresses in a number of polymer processing operations is described.

The deformation-induced internal stress can be investigated by stress relaxation procedures, and has been found to be sensitive to the state of ageing or annealing, although no quantitative interpretation has been developed to date.

The layer removal process for determining residual stress distribution is described in some detail and examples of applications are presented, including a full biaxial treatment. The results of this technique are quite revealing and some of the practical implications are indicated.

1. INTRODUCTION

There are numerous references in the literature to the development and presence of internal stresses within moulded plastics and to their influence on properties. The author has made several contributions, concentrating primarily on the study of injection mouldings, and a short review, on the assessment of internal stresses in injection-moulded thermoplastics in which the experience of this laboratory

165

Polymer Testing 0142-9418/84/$03·00 © Elsevier Applied Science Publishers Ltd, England, 1984. Printed in Northern Ireland

has been described, appeared recently.[1] Much of the detail presented in Reference 1 has been omitted from the present paper, and emphasis is placed on the most recent developments.

It is necessary first to define what is meant by 'internal stress' for this term is used in the literature for two quite different physical phenomena and is often misused for a third. The uses which, in the opinion of the author, are legitimate, are as follows:

1. 'Internal stress' is often used to describe the presence in a body of a self-stressed state. The local state of stress varies from point to point in the moulding, even when it is free from surface tractions, and the shape taken up by the body will depend on the geometry of the stress distribution. Stresses of this kind develop if temperature gradients exist while a moulding sets and are therefore often simply identified as 'moulding stresses'. A preferred term for this class of phenomena is 'residual stress', which also embraces curing stresses in thermosets. The term 'residual stress' is adopted in this paper.

2. 'Internal stress' is often used in the field of physical metallurgy to indicate a resistance to deformation, and is sometimes described instead as a 'back stress' or 'friction stress'. The behaviour of polymers is phenomenologically similar to that of metals (although the mechanism responsible is almost certainly different) and a number of studies have been made.[2-5] This phenomenon may be referred to as 'deformation-induced internal stress'. It is equal to that part of the applied stress which does not disappear as a consequence of non-elastic processes, for example in a stress relaxation test it will equal the ultimate stress at infinite time.

The two phenomena are interactive and both are dependent on the conditions of fabrication and it is always necessary to interpret the results of experiments designed to measure either one most carefully.

The third use of the term 'internal stress' to be found in the literature is, in the opinion of the author, misleading, and is in the description of a state of frozen-in molecular orientation. A degree of molecular orientation is retained in many plastics articles. In those formed by moulding operations involving melt flow (extrusion, injection moulding, blow moulding), molecules are oriented in the melt and cooling usually takes place too rapidly to permit restoration of random conformation. In solid forming operations, such as fibre drawing and thermoforming, molecular orientation is again produced and there is very little opportunity for subsequent recovery. Although it is true that an article containing oriented molecules would generate within itself a stress if heated above its glass transition temperature while its dimensions were kept fixed, the author does not approve of the use of the term 'internal stress' for this characteristic and prefers instead Andrews' term[6] of 'passive stress'. It is acknowledged that it will be difficult to isolate the effect of molecular orientation on any property from the effects associated with residual stress and it is further expected that these phenomena will be interactive.

2.   ORIGINS OF INTERNAL STRESSES

2.1. Residual stresses

In most plastics moulding operations a temperature gradient develops during cooling and the material solidifies progressively as the locus of the solidification temperature travels inwards from the cool surface. The severity of the temperature gradient depends on the type of moulding operation, the geometry of the article (especially the section thickness) and the operating conditions, which may or may not include forced cooling. In extrusion, the extrudate may cool slowly in stagnant air at one extreme, or rapidly by feeding it into a cold-water bath. With injection moulding the mould cavity wall will be held at a temperature below the solidification temperature (but often well above ambient) and the moulding cools by conduction, being in contact with a large heat sink (the mould). In blow moulding, cooling begins in the mould, although in this case only one surface is in contact with the mould giving rise to thermal conditions quite different to those obtained with injection moulding. Forced cooling can be achieved by spraying cold vapour into the moulding.

It is convenient to consider first the changes which take place within an injection moulding during cooling. The material which finds itself adjacent to the mould cavity wall sets rapidly to form a skin which, at the solidification temperature, has the dimensions of the cavity. This skin is a poor thermal conductor and prevents the material in the interior from setting rapidly. This will permit any molecular orientation which developed during the mould-filling stage to relax either partially or fully. For simplicity, consider that the gate freezes off fairly early on so that no more material can be delivered to the cavity from the injection system. Thus as the material continues to cool and to undergo thermal contraction the material in the interior (core) will attempt to shrink, a process opposed by the skin which has already solidified. The material in the interior hence goes into a state of hydrostatic tension. This may be partly relieved by the formation of sink marks or, in serious cases, internal voiding. Sinking is caused by the component of stress acting normal to the surface in which it appears. Stresses in directions parallel to the surface are less easily relieved and it is these which are measured by the layer removal technique referred to later. If there is no influence of flow then these stresses parallel to the surface would be expected to be equibiaxial.

In the case of an extrudate, biaxial stresses will again be found in the surface. An important case is that of extruded pipe. If this is cooled simply from the outer surface then the inner surface may solidify last and so develop tensile stresses. Since flaws are found at the surface of extrudates more often than in the interior it is most disadvantageous to have tensile residual stresses at the surface. If, instead, compressive stresses are present the critical applied stress causing fracture from such a flaw will be increased, and it is therefore desirable to manufacture pipes with compressive stresses at both the inner and the outer surfaces. This can be achieved by applying forced cooling to the inside as well as to the outside of the pipe.

Several theoretical analyses of the development of residual stresses in plastics mouldings have been made. The solidification process is most complex. The relaxation behaviour is sensitive to temperature but changes progressively, not discontinuously, as the material cools, and it is difficult to take into account the recovery of flow-induced molecular orientation. A further complication is the special molecular reorganisation demanded by the formation of crystals during the solidification of crystallising polymers. In the simplest analyses these effects are largely neglected, and a first approximation is obtained by considering that the material has a discrete solidification temperature at which the properties change discontinuously from those of a fluid to those of a solid, and that the solid responds in a linear elastic fashion to the thermal contraction. This leads to the prediction that the through thickness residual stress distribution will be parabolic for a moulding with rectangular cross-section if it is cooled equally on both faces. This profile will be approximately valid for a flat-faced injection moulding or for a thin-walled pipe, cooled equally internally and externally.

The profile predicted by analyses in which the viscoelastic behaviour of the material is taken into account is less simple; the central region of the parabola (the tensile interior region) becomes flatter and the stress near to the surface is quite sensitive to the model used, but there is normally a steep stress gradient in the vicinity of the point where stresses reverse from compressive to tensile as is true in the parabolic approximation. Some of the trends indicated by these theoretical studies have been observed in experimental investigations, but the measured profiles obtained with crystallising polymers and polymers containing short fibre fillers are very complicated and no firm theoretical formulation has yet been developed for these materials. A selection of references dealing with the theoretical prediction of residual stress distributions in flat plates[7-13] and in pipes[14] is included here, and further papers are cited therein.

2.2. Deformation-induced internal stresses

The origins of this class of internal stress have been studied in depth in metals, as can be confirmed by referring to two authoritative reviews.[15,16] The mechanisms which account for the deformation-induced internal stresses in metals involve the motion and interaction of dislocations and cannot be invoked to explain observations made on non-crystalline polymers in which the behaviour is phenomenologically similar. Instead the site model theory of deformation has been developed to model the behaviour under various loading conditions.[17-20]

The site model theory is based on the supposition that a segment of a molecule can exist in two isomeric forms, shown schematically in Fig. 1. Thermally activated transformations from one state to the other are occurring all the time, and the frequency of transformation in any particular direction will depend on the energy barrier. If a particular set of conditions (including stress and temperature) is

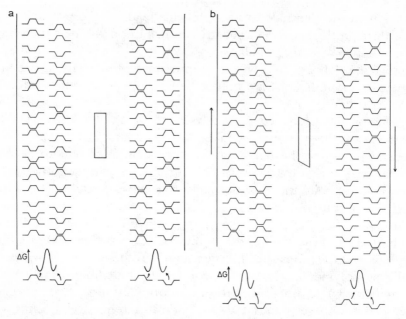

Fig. 1. (a) Schematic diagram of segments of molecules in an unstressed bar with site populations chosen by the toss of a coin (i.e. it is assumed that both sites are equally probable in the unstressed condition, as shown by the free energy diagram at the bottom). (b) The modification to site populations caused by an applied shear stress. Sites marked ⌒⌒ on the left-hand side and ⌒⌒ on the right-hand side in (a) have been decided again on the toss of a coin, while those marked ⌣⌣ on the left-hand side and ⌒⌒ on the right-hand side have been kept the same. The energy level diagrams have been amended to indicate the preferred conformation in accordance with this scheme. The macroscopic dimensions reflect these changes, and are represented by the inset figures in the centre.

maintained for long enough a dynamic equilibrium will be established in which the number of site transformations in one direction will equal that in the opposite direction. The form with the lower energy will be more heavily populated and the population partition, N_1/N_2, will be given by:

$$N_1 \exp\left(-\Delta G_1/kT\right) = N_2 \exp\left(-\Delta G_2/kT\right) \qquad (1)$$

where ΔG_1 is the height of the activation barrier for transformation in the direction site 1 to site 2, k is the Boltzmann constant, T is the absolute temperature, and ΔG_2 is the value for transformation in the opposite direction. For simplicity, Fig. 1(a) has been constructed for the case in which $\Delta G_1 = \Delta G_2$ in the unstressed state, so that each segment will take the forms site 1 and site 2 with equal probability. The selections shown in Fig. 1(a) were chosen by the toss of a coin. When a stress is applied a bias will be applied to the jumping direction. One of the isomeric states will have its activation barrier increased while the other will suffer a decrease. As a consequence there will be a greater number of jumps in the stress-aided direction until dynamic equilibrium is restored with a new population distribution. This is depicted in Fig. 1(b) in which the new population has been chosen in the following way. All segments which in the unstressed state shown in Fig. 1(a) have the orientation which becomes stress-favoured are re-plotted unchanged in Fig. 1(b). All segments which in Fig. 1(a) have the other orientation have had their orientation chosen again by the toss of a coin before plotting in Fig. 1(b). Movements of this kind occurring throughout the body of the material will, of course, be reflected in the macroscopic dimensions.

If the body is loaded by the application of a fixed deformation (stress relaxation test) the stress will decay during the period in which the number of jumps in the stress-aided direction exceeds the number in the opposite direction, but will reach a constant value when a dynamic site population equilibrium distribution is re-established. This constant stress is identified with the deformation-induced internal stress.

The mathematical analysis of the two-site theory has been dealt with elsewhere[17-20] and has been shown to correspond to the behaviour of a standard linear solid.

It is worth remembering that residual stresses set up during moulding will partially relax, causing a population redistribution. The residual stresses must eventually equilibrate so that at all points the local residual stress will be equal to the internal stress generated by the local site population imbalance.

3. MEASUREMENT OF INTERNAL STRESSES

3.1. General survey of residual stress measurement

Many of the methods of measurement of internal stresses in plastics were developed first for metals, but the properties of the two classes of material are very different and some techniques have proved able to adapt to the requirements of the study of plastics more than others. A popular technique for measuring residual stress in metals is to fix a system of strain gauges to the surface of a component suspected to contain residual stress, and then to drill a hole in the component nearby and measure the strains that are produced at the surface when the consequent relaxation takes place. A standard array of strain gauges is normally chosen for this purpose, with a centrally positioned standard-sized hole. There are very few reports of successful application of this technique to polymers. The drawbacks with the technique include the problem of finding a suitable adhesive to attach the strain gauges to the polymer surface, and the difficulty in drilling a reproducible hole in polymeric samples without introducing a stress field by the machining process. Attempts to use this class of technique have been reviewed by Ito.[21] A related technique involves the removal of thin layers of uniform thickness from the surface of the moulding and measuring the resulting changes in dimension. This technique can be applied conveniently only to parallel-sided articles with planar or cylindrical faces because of the difficulty in removing uniform layers from other shapes. Variations of this technique have been employed in several laboratories and it is described more fully in Section 3.2.

Another characteristic which can provide information about residual stress is optical birefringence. The problem with this technique is that birefringence has a number of causes. The polarisability of chemical bonds changes when they are stressed, giving rise to the photo-elastic response upon which measurements of residual stresses might be based. On the other hand, chemical bonds are directional and in a non-isotropic material the presence of favoured bond orientations will produce birefringence, so that in an injection-moulded article containing frozen-in molecular orientation this may provide a much larger contribution than the residual stresses. If the temperature of the article is well below the glass transition temperature then this orientation will be stable and it is easy to establish which contribution is dominant by making a saw-cut into the article and inspecting it between crossed polars. If residual stress is the principal cause of birfringence then relaxation associated with the saw-cut and the newly formed surfaces will be revealed by a perturbation in the isochromatic fringe pattern. On the other hand, if the birefringence is caused by orientation the fringe patterns will be continuous and unperturbed across the saw-cut.[1,22] When residual stress and orientation are significant it is difficult to separate their contributions, although this problem has been tackled with notable success by Saffell and Windle.[23]

A further point to consider is that many polymers possess highly polarisable side-groups and that the orientation of these will be determined only partly by the main chain conformation. As an example, in Fig. 2 the orientation of the bond marked (a) connecting a phenyl residue to the backbone in a polystyrene molecule which has locally an all-*trans* sequence will be largely decided by the orientation of the backbone, but rotation about bond (a) will be determined as much by intermolecular as by intramolecular forces and may be influenced by the state of stress. This contribution is in kind more like the orientation contribution already mentioned, but does not give any direct information about molecular backbone conformation which, because of its influence on mechanical properties and dimensional stability, is usually the characteristic of interest. On the other hand, the side-group contribution is often easily influenced by changes in stress and it is difficult to extract meaningful conclusions when this contribution is significant.[24,25] The method of analysis presented by Read[26] at this meeting indicated that the birefringence probably contained a contribution which was neither totally orientation-dependent nor totally stress-dependent, and may be related to this phenomenon.

Another quantitative method which has been described for residual stress measurement is that of Barrett and Predecki.[27,28] They introduced small crystalline particles into their moulding resins and deduced stress levels from x-ray lattice parameter measurements. This technique would appear to be better suited to studies of thermosets

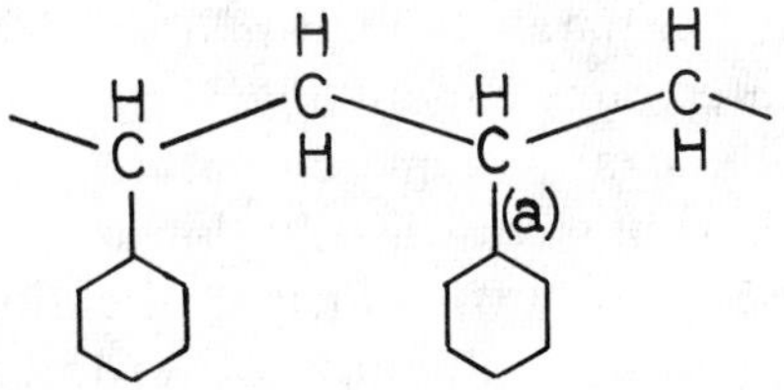

Fig. 2. A short all-*trans* sequence in polystyrene.

in casting and hand lay-up articles than for moulding operations involving flow, especially with crystallising thermoplastics.[1]

In addition to the quantitative methods for residual stress analysis, a number of techniques exist which may reveal their presence without indicating their magnitude. Each of the techniques listed below is sensitive to molecular orientation as well as to residual stress, and this is probably the main barrier to quantifying the method:

1. Distortion at elevated temperature; this will be more and more a consequence of molecular orientation as the temperature is increased.
2. Craze patterns promoted in mouldings containing residual stresses when submerged in an aggressive medium.
3. Directional damage during scanning electron microscope inspection.[29]

Another technique worth adding to this list is the use of controlled indentation in which the pattern of deformation (crazes, etc.) around the indentation identifies directions of weakness;[30,31] these will be a consequence of the local molecular orientation and residual stresses and although this method may be the basis of a useful practical tool it does not lend itself to a direct assessment of residual stress level.

In the following section the layer removal procedure is dealt with in more detail. This is the technique which produces information with the most unambiguous interpretation and is the one which has received the most attention in the study of residual stresses in polymer mouldings.

3.2. The layer removal procedure

The analysis employed most frequently in the study of residual stresses in polymeric articles is that introduced by Treuting and Read[32] in which uniform thin layers are removed from mouldings in the shape of rectangular parallelipipeds. After each layer removal the curvature of the unrestrained moulding is measured. If the stresses are uniaxial or equibiaxial a single curvature measurement at each stage suffices. If a biaxial stress distribution is present the curvature should be measured in the two principal directions (e.g. along flow and perpendicular to flow in the case of an injection-moulded plaque). Methods of measuring curvature are described in Reference 1

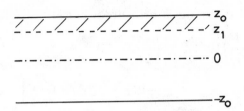

Fig. 3. Cross-section of a flat parallel-sided moulding showing the mid-plane prior to machining at $z = 0$, the as-moulded surfaces at $z = \pm z_0$ and the surface after machining away layers from the upper half of the bar at $z = z_1$.

and summarised below. The next step is to plot values of curvature, ρ, against the depth of material removed, $(z_0 - z_1)$, see Fig. 3. Graphs for the curvature parallel and perpendicular to flow must be obtained separately for the biaxial case. The ρ versus $(z_0 - z_1)$ plots are then converted to a stress versus $(z_0 - z_1)$ profile using the formula presented by Treuting and Read:[32]

$$\sigma_{i,x}(z_1) = \frac{-E}{6(1-v^2)}\left[(z_0 + z_1)^2\left\{\frac{d\rho_x(z_1)}{dz_1} + \frac{v\,d\rho_y(z_1)}{dz_1}\right\}\right.$$

$$\left. + 4(z_0 + z_1)\{\rho_x(z_1) + v\rho_y(z_1)\} - 2\int_{z_1}^{z_0}\{\rho_x(z) + v\rho_y(z)\}\,dz\right] \quad (2a)$$

where $\sigma_{i,x}(z_1)$ is the residual stress in the x direction at the plane z_1 from the mid-plane of the specimen prior to layer removal. E, the Young's modulus, and v, the Poisson's ratio, are both assumed to be isotropic in the x–y plane, and uniform throughout. ρ_x and ρ_y are the components of curvature in the x- and y-directions, respectively.

3.2.1. *Special cases*
Three special cases merit attention:

1. Firstly, if the stresses are uniaxial, i.e. $\sigma_{i,y} = 0$, then from eqn. (2a):

$$\sigma_{i,y}(z_1) = 0 = \frac{-E}{6(1-v^2)}\left[(z_0 + z_1)^2\left\{\frac{d\rho_y(z_1)}{dz_1} + \frac{v\,d\rho_x(z_1)}{dz_1}\right\}\right.$$

$$\left. + 4(z_0 + z_1)\{\rho_y(z_1) + v\rho_x(z_1)\} - 2\int_{z_1}^{z_0}\{\rho_y(z) + v\rho_x(z)\}\,dz\right] \quad (2b)$$

and it follows that $\rho_y(z_1) = -v\rho_x(z_1)$. Hence if the curvature ρ_x

in the axial direction is measured, eqn. (2a) becomes

$$\sigma_{i,x}(z_1) = \frac{-E}{6} \left[(z_0 + z_1)^2 \frac{d\rho_x(z_1)}{dz_1} \right.$$

$$\left. + 4(z_0 + z_1)\rho_x(z_1) - 2 \int_{z_1}^{z_0} \rho_x(z)\, dz \right] \quad (2c)$$

2. If the curvature after layer removal is very small in one direction, (i.e. $\rho_y = 0$) then eqn. (2a) becomes:

$$\sigma_{i,x}(z_1) = \frac{-E}{6(1-\nu^2)} \left[(z_0 + z_1)^2 \frac{d\rho_x(z_1)}{dz_1} \right.$$

$$\left. + 4(z_0 + z_1)\rho_x(z_1) - 2 \int_{z_1}^{z_0} \rho_x(z)\, dz \right] \quad (2d)$$

and $\sigma_{i,y}(z_1) = \nu\sigma_{i,x}(z_1)$.

Equation (2d) has been used in several investigations.[33-5]

3. If the stresses are equibiaxial then $\rho_x(z_1) = \rho_y(z_1) = \rho$ (say), and:

$$\sigma_{i,x}(z_1) = \sigma_{i,y}(z_1) = \frac{-E}{6(1-\nu)} \left[(z_0 + z_1)^2 \frac{d\rho}{dz_1} \right.$$

$$\left. + 4(z_0 + z_1)\rho - 2 \int_{z_1}^{z_0} \rho\, dz \right] \quad (2e)$$

It is apparent that eqns. (2c), (2d) and (2e) differ only in the value of the factor outside of the square bracket on the right-hand side.

3.2.2. Method of layer removal

In the case of mouldings in the form of rectangular parallelipipeds uniform thin layers are removed by milling using a high cutting speed to minimise the energy dissipated during the fracture and deformation of the specimen. In a series of experiments conducted by Hindle[36] three different cutting arrangements were compared, each having the specimen mounted horizontally on the machine bed. It is sufficient to mount the specimen using double-sided adhesive tape which dispenses with the need for clamps and prevents the specimen from bending in response to the unbalanced forces which develop during this procedure. The whole of the surface of a bar can be machined in a single sweep. The cutters tested were: (1) a 'side and face cutter' milling tool with its axis horizontal; (2) a four-fluted slot-cutting end mill with its axis vertical; and (3) a single point cutter mounted on the milling machine with the point rotating in a horizontal plane. The most satisfactory cutting was obtained with the single point cutter, whereas the four-fluted end mill caused melting and was not used in any of the experiments for which we have reported results. Machined surfaces were inspected in the scanning electron microscope and no evidence for melting or any other kind of machining damage could be found for samples cut using the single point cutting tchnique.

Cooling the specimen surface by directing a stream of liquid nitrogen onto it for about 2 min prior to a milling operation and throughout its execution did not cause any detectable change with polystyrene specimens, but a small yet significant effect was observed in the curvature versus depth-removed profiles for injection-moulded glass-fibre-filled polypropylene.

3.2.3. *Measurement of curvature*

For bar-shaped specimens a common method of curvature measurement is to place a bar concave downwards onto a plane surface and to measure the height of the bar centre above the surface. If, as is common, the height is measured by advancing a probe with a micrometer drive until it contacts the bar surface then interference with the probe is possible. This method has other shortcomings[1] and these problems were avoided by using, instead, a non-contact optical method in which the angle of deflection of a laser beam from mirrors attached to the bar surface is used.[1,34] With this technique as with the micrometer probe methods the analysis makes the implicit assumption that the bar curves uniformly, but this need not necessarily happen. To check the uniformity of curvature Hindle[36] constructed a series of circular arcs of known radius and compared machined bars directly by eye. For the vast majority of the bars tested it was found that no variation in curvature could be detected. This procedure could be used for curvature measurement and experience leads to the belief that acceptable accuracy could be achieved if the following recommendations were heeded. Firstly, the arcs should be inscribed on a suitably permanent surface. Secondly, a holder should be constructed in which a bar could be held at its centre with line contact (as for the laser reflection technique) and which would hold the bar close to, but not in contact with, the surface containing the comparison curves.

With plaques it is important to measure curvature in more than one direction. Even with small mouldings ($\sim$100 mm diameter) the warping is very complicated and curvature changes continuously over the whole surface.[37-9] Although optical techniques, such as the shadow moiré method, give a very clear indication of the distortion of these mouldings they do not readily lend themselves to curvature measurement in the presence of such variations. Furthermore, it is much more difficult to mill away uniform layers from a plaque because (1) the use of double-sided tape to attach the plaque to the bed is unlikely to be adequate in many cases, and (2) single sweep removal from the whole surface may not be possible. The approach used in the author's laboratory has been to machine bar-shaped specimens from plaques and conduct the layer removal procedure on them. They are generally found to have acceptably uniform curvature and the analysis has been conducted in the conventional way. By cutting bars both parallel and perpendicular to flow, curvature measurements have been obtained which can be used to represent $\rho_x(z_1)$ and $\rho_y(z_1)$ in eqn. (2a), so permitting a full biaxial analysis. Although creation of new surfaces (on extracting the bars from the plaques) will modify somewhat the stress distribution, this technique appears to be the most useful solution to the problem of assessing stresses in plaques, for which none of the special cases listed in Section 3.2.1 apply.

3.2.4. *Direct measurement of bending moment*

In the layer removal procedure presented above, the curvature of a bar containing unbalanced stresses is used to estimate the bending moment associated with the layer which has been removed to create the imbalance and hence to find the stress it contained prior to machining. The Treuting and Read derivation is based on an elastic analysis and it is not easily modified for bars in which the stiffness or Poisson's ratio changes through the depth. An alternative experimental procedure is to measure the bending moment required to straighten the bar at each layer removal step.[1] In this way it is much easier to introduce a modification to account for a depth-dependent modulus. Attempts to conduct the analysis by this procedure have been unsuccessful because of irreproducibility, although the magnitudes of stress measured were consistent with those obtained by the more common form of the analysis. There is considerable scope for improving the measuring method and this procedure merits re-examination.

3.3. Measurement of deformation-induced internal stress

In methods for measuring deformation-induced internal stresses, the assumption is normally made that this internal stress opposes the applied stress and that in cases such as uniaxial loading or simple shear loading in which the applied stress is uniform $(=\sigma)$ the deformation-induced internal stress is also uniform $(=\sigma_i)$. It is then proposed that the deformation kinetics should be described in terms of an 'effective stress' $=(\sigma - \sigma_i)$ rather than σ. The deformation mode for which most information is available on polymers is uniaxial extension and a method for deriving the internal stress has been developed by Li,[40] working with metals.

The method used by Li[40] starts with the premise that stress relaxation follows a power law relationship such that:

$$\frac{d\sigma}{dt} = -EB(\sigma - \sigma_i)^n \tag{3}$$

where E is the Young's modulus, and B and n are material parameters. This is consistent with the definition of σ_i as the ultimate stress after a very long period at constant deformation, for eventually $d\sigma/dt = 0$. To conduct a Li analysis the stress relaxation data are plotted as σ versus $\log t$, giving a sigmoidal curve (Fig. 4), and the gradient of this curve is measured for various values of σ less than that at which the point of inflection occurs. If this gradient is plotted against stress an approximately straight line is obtained, and this is extrapolated to meet the σ axis. At this point $d\sigma/d \ln t = 0$ so that the intercept can be taken to be $\sigma = \sigma_i$. This should be the result which would be obtained by prolonging the measurement of stress relaxation until no more change is detected.

If the material does not obey the power law of stress relaxation then a straight line extrapolation need not be obtained in a Li plot, although it would still be true that the ultimate (σ_i) value would be obtained at the point of intersection with the σ axis if a reliable curved extrapolation could be made. If the material follows, instead,

J. R. White

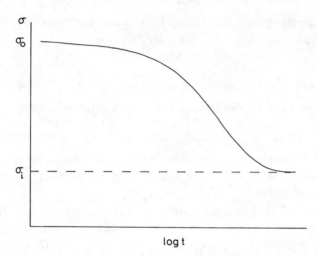

Fig. 4. Schematic representation of stress relaxation plotted against logarithm of time. σ_0 is the initial stress and σ_i the value approached at long times.

the deformation kinetics predicted by the site model theory then a plot of $1/t(d\sigma/d\ln t)$ versus σ should give a straight line, facilitating extrapolation, and the intercept with the σ axis can again be identified with σ_i.[20,41]

Whatever the method of extrapolation, the result obtained may be influenced by residual stresses. One way of testing this is to perform a series of measurements at different deformation levels, giving a series of values for σ_i, each with a corresponding value for σ_0, the initial stress. By plotting σ_i versus σ_0 and extrapolating to zero σ_0 (at which the deformation-induced component must be zero), it can be shown whether or not a second component (caused by the presence of residual stress) exists.[35]

If by this procedure or by the Kubát and Rigdahl procedure (described in Section 3.4) a non-zero residual stress component is found to be present in σ_i, then an explanation is demanded since the net stress across a section of a self-stressed body (free from external tractions) should be zero. Kubát and Rigdahl[42] have shown that if the relaxation behaviour of material at different levels in a moulding is different then a non-zero contribution to σ_i will be produced if there is a residual stress distribution. It is well known that polymeric mouldings, especially those produced by injection moulding, contain a skin-core morphology with two or more regions with distinctly different structures, and large differences in behaviour are expected at different depths.[43-57]

3.4. Kubát and Rigdahl analysis

The Kubát and Rigdahl (KR) analysis cannot easily be categorised as a method to obtain information about residual stress distribution or the magnitude of the deformation-induced stress. Instead it produces an 'internal stress parameter' which is, in some ways, a hybrid between the two.

To conduct a KR analysis[2,42] a series of stress relaxation tests at different deformation levels must be performed on nominally identical specimens and the data plotted as σ versus $\ln t$. Kubát and Rigdahl favour the power law description of relaxation behaviour and manipulation of eqn. (3) shows that if σ_i is treated as a constant, then a plot of $(-d\sigma/d \ln t)_{max}$ (the gradient of the σ versus $\ln t$ curve at the point of inflection) versus the initial stress, σ_0, for a series of specimens should be a straight line. Closer examination shows that this is true also if:

$$\sigma_i = \sigma_{i,r} + \sigma_{i,d} = \sigma_{i,r} + p\sigma_0 \qquad (4)$$

where σ_i has been decomposed into a residual stress component, $\sigma_{i,r}$, and a deformation-induced component, $\sigma_{i,d}$, which has further been assumed to be proportional to the initial stress, and p is a constant. This step is not explained explicitly in the publications by Kubát and Rigdahl.

If the straight line is extrapolated to meet the σ_0 axis this corresponds to the condition $\sigma_i = \sigma_0$, and this value is taken to be the 'internal stress parameter'. Kubát and Rigdahl[42] show that this should be zero if a specimen contains no residual stresses. If, as Kubát and Rigdahl imply, their extrapolation is intended to isolate $\sigma_{i,r}$, then it should be noted that rearrangement of eqn. (4) after substituting $\sigma_i = \sigma_0$ leads to:

$$\sigma_{i,r} = (1 - p)\sigma_0 \qquad (5)$$

If the material truly follows the power law, then $\sigma_{i,r}$ should equal the intercept on the σ_i versus σ_0 plot obtained from a series of Li analyses, as described in Section 3.3, and p should be the gradient of such a plot.

It has already been pointed out that the average value of the residual stress across a section must be zero, and to record a non-zero value of $\sigma_{i,r}$ by the KR procedure requires the simultaneous fulfilment of two conditions: (1) the residual stresses must be non-zero, and (2) there must be a variation in the relaxation behaviour of the material at different positions within the body.[42] Such variations will be dependent on the morphology and on the state of ageing of the material, both of which are dependent on the thermo-mechanical history of the material just prior to, during, and after solidification and will therefore be sensitive to processing conditions and to any post-moulding conditioning such as annealing. Since the residual stress distribution is likewise controlled by the processing conditions and post-moulding conditioning, it is difficult to see exactly how to separate the causes of any measured variations in $\sigma_{i,r}$ found by KR analysis.

4.   RESULTS AND DISCUSSION

The techniques described above have been used to examine the effect on injection mouldings of production conditions,[58] annealing with a uniform temperature[41,59] or in a temperature gradient,[37,60] ageing[34,38] weathering,[61] crazing,[58] and cross-linking[35] variously in crystallising,[34,35,60,61] non-crystallising,[58-61] and short glass-fibre-filled thermoplastics.[37,38,60] A preliminary study of an epoxy thermosetting resin has also been conducted.[62] These studies are reported in detail elsewhere and it is not intended to review the results again, only to draw on examples to illustrate specific points.

4.1. Layer removal analyses

4.1.1. *General remarks*
In the majority of injection mouldings it has been confirmed that the residual stress profile is tensile in the interior and compressive near to and at the surface. However, some recent work has shown that it is possible to reverse the sense of the stress distribution, and examples in which tensile stresses were generated at the surface have been found by Sandilands[63] using certain storage conditions, by Thompson[60] when annealing specimens in a temperature gradient, and by Qayyum and White[61] investigating specimens weathered in a very hot climate.

Although as-moulded bars have always been found to have compressive stresses at the surface and tensile stresses in the interior, the predicted parabolic distribution has rarely been indicated. In order to record a parabolic distribution the ρ versus $(z_0 - z_1)$ plot should be a straight line as is easily confirmed by substituting:

$$\rho = -a(z_0 - z_1) \tag{6}$$

into eqn. (2c) whence the result becomes:

$$\sigma_{i,x}(z_1) = \frac{Ea}{3}(z_0^2 - 3z_1^2) \tag{7}$$

with similar results if eqns. (2d) or (2e) are used instead. In eqn. (6) a is a constant, and a negative sign is used if the bar bends to become concave; in most previous publications the sign of ρ has been reversed for ease of presentation, and straight-line plots of ρ versus $(z_0 - z_1)$ have been the exception, although examples found with polystyrene are reported in Reference 58. Once the curvature versus depth removed plot departs from linearity it becomes necessary to conduct the full term-by-term Treuting and Read analysis, and the stress profile is no longer parabolic. In general, the departure from parabolic is more pronounced with crystallising polymers than with non-crystallising polymers,[34,38] and the residual stress profile becomes especially complicated with glass-fibre-filled polypropylene.[38]

4.1.2. *Biaxial analysis*

Another feature of the work conducted with glass-fibre-filled polypropylene is the marked anisotropy of the specimens. The specimens used for this study were injection-moulded plaques and Thomas *et al.*[39] have shown that these mouldings contain a markedly non-isotropic fibre orientation distribution and that the modulus (by ultrasonic measurement) is anisotropic. The residual stress distributions, assessed using eqn. (2d), were fairly complicated both for bars cut parallel to the flow direction and perpendicular to it (Fig. 5). Inspection of the data indicates that the assumption implicit in the computation of stress perpendicular to the flow direction ($\sigma'_{i,y}$ in Fig. 5(b)), that ρ_x is negligible, is a very poor one. This is clearly a case where the full biaxial treatment should be applied. Inspection of eqns. (2a) and (2d) shows that this can very easily be completed, for if the stress distributions computed from eqn. (2d) are $\sigma'_{i,x}$ and $\sigma'_{i,y}$,

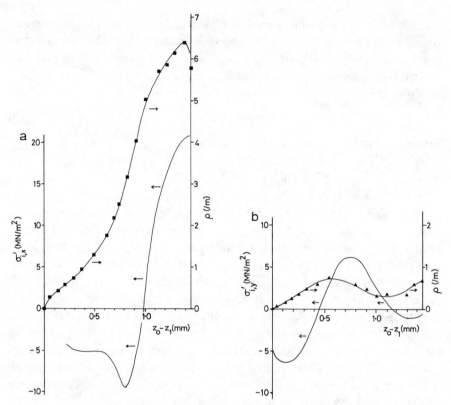

Fig. 5. (a) Plot of curvature, ρ, versus amount of material removed, $(z_0 - z_1)$, for a bar cut parallel to the flow direction from a glass-fibre-filled polypropylene injection-moulded plaque. The stress distribution, $\sigma'_{i,x}$, was computed using eqn. (2d). (b) Results from a bar cut transverse to the flow direction from a plaque moulded under identical conditions to that used in Fig. 5(a). The stress distribution, $\sigma'_{i,y}$, was computed using eqn. (2d). Further discussion of the work on glass-fibre-filled mouldings from which these results are taken can be found elsewhere.[38] The bar centre coincides with the right-hand axis.

then the full biaxial formula, eqn. (2a), gives:

$$\sigma_{i,x} = \sigma'_{i,x} + \nu\sigma'_{i,y} \tag{8a}$$

and

$$\sigma_{i,y} = \sigma'_{i,y} + \nu\sigma'_{i,x} \tag{8b}$$

It is evident that in the example cited here the correction to $\sigma'_{i,x}$ to give $\sigma_{i,x}$ is significant, yet does not dominate, whereas the term $\nu\sigma'_{i,x}$

is the dominant term on the right-hand side of eqn. (8b). This is confirmed in the plots of $\sigma_{i,x}$ and $\sigma_{i,y}$ (Fig. 6).

I believe that this is the first example to be published of the full biaxial Treuting and Read analysis with a polymer moulding, and it illustrates the importance of using it when the stresses are markedly anisotropic. The character of the computed stress distribution for the dominant direction is not much altered, but in the orthogonal direction (across the flow in the example described here), the computed level of stress is found to be totally changed when applying the full analysis. The procedure described above in which the quantities $\sigma'_{i,x}$ and $\sigma'_{i,y}$ were computed initially from the corresponding curvatures, ρ_x and ρ_y, is quite convenient, for each of the measurements (of $d\rho/dz_1$ and $\int \rho \, dz$) required to compute the intermediate distributions, $\sigma'_{i,x}(z_1)$ and $\sigma'_{i,y}(z_1)$, would be required for a direct Treuting and Read analysis, and it is a simple matter to use eqns. (8a) and (8b) to complete the computation of $\sigma_{i,x}$ and $\sigma_{i,y}$.

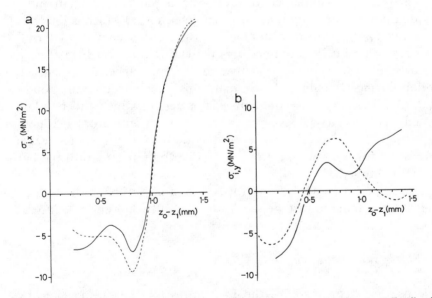

Fig. 6. (a) Biaxial Treuting and Read analysis for the residual stress distribution, $\sigma_{i,x}$, in the flow direction of the plaques used for Fig. 5; $\sigma'_{i,x}$ from Fig. 5(a) has been re-plotted for comparison (broken line). (b) Biaxial Treuting and Read analysis for the residual stress distribution. $\sigma_{i,y}$, across the flow direction of the plaques used for Fig. 5; $\sigma'_{i,y}$, from Fig. 5(b) has been re-plotted for comparison (broken line).

4.1.3. *Non-symmetric stress distributions*

In many cases it is expected that the stress distribution within a moulding will be symmetric about its mid-plane. Thus all that is required is to perform an analysis in which layers are removed down to this level. Removing further layers should confirm the validity of the assumption, but for thin specimens this leads to practical difficulties because of the fragile nature of the remaining material. Thus, to check on the symmetry of the distribution, it is necessary to take a second, nominally identical, specimen and to remove layers from the opposite side.

The investigation in which this procedure has been most used to date has been one in which the influence of annealing moulded bars in a temperature gradient has been studied.[60]

After annealing in a temperature gradient the bars become bent. Thus the analysis starts with the measurement of the initial curvature before layers are removed (a procedure advised for all layer removal tests, even if asymmetry is not expected). In the subsequent computation of stress, the initial curvature value is subtracted from $\rho(z_1)$ which has an influence on both the term in eqn. (2a) which is linear in ρ and the $\int \rho \, dz$ term (but not on the differential coefficient term). The stress distribution obtained in this way is that which resides in the bar after it has bent in response to the imbalanced forces which develop during annealing. To estimate this imbalanced component a standard elastic analysis could be performed in which the bending moment required to achieve the observed initial curvature is found. This component would have to be added to the residual stress distribution for the bent bar to give the distribution present if the bar were straightened.

An example of an analysis performed on a pair of bars 3-mm thick annealed in a temperature gradient for 12 min by holding one face at 90 °C and the other at 25 °C is shown in Fig. 7.

4.2. Stress relaxation

Little use has been made of the Li analysis for the reason cited by Kubát and Rigdahl, namely the problem of measuring with sufficient accuracy the very slowly changing gradient of the σ versus $\ln t$ curves. Although in limited studies some straight-line Li plots have been found,[35,62] other examples have also been found in which the

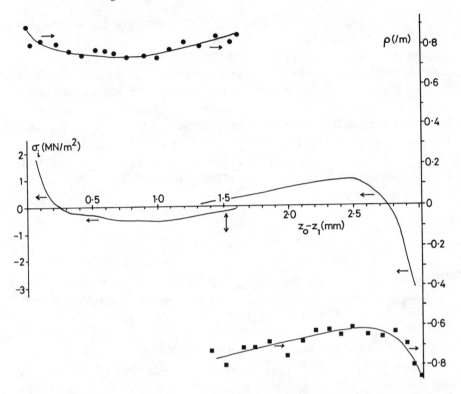

Fig. 7. Layer removal analysis of a pair of injection-moulded polystyrene bars heat-treated together for 12 min in a temperature gradient with hot face/cold face combination 90/25 °C. The results obtained with the bar from which layers were removed from the hot face are shown on the left and those obtained with the bar machined on the cold face are shown on the right. The right-hand axis (ρ) coincides with the cold face, ($z_0 - z_1$) values are measured from the hot face, and the bar centres coincide at the position marked with the vertical arrow. The hot face becomes concave after such a treatment, and its curvature is shown as positive. Further discussion of gradient annealing can be found elsewhere.[60]

plot is curved. The alternative method for finding σ_i, based on the site model theory, in which a plot of $1/t(d\sigma/d \ln t)$ versus σ is proposed, has also shown a curved relationship.[41] These plots curve in the opposite sense to the Li plots based on the same data, but the extrapolated values of σ_i seem to be in agreement.[41] This should, of course, be so since the extrapolation is effectively locating the stress limit at infinite time in both cases.

Many examples have been presented of the use of the Kubát and Rigdahl procedure[35,41,58,62] and it has always been found that the $(-d\sigma/d \ln t)_{max}$ versus σ_i plot gives a good approximation to a straight line, as predicted. No uniform pattern to the values of σ_i obtained has been perceived, however, although certain trends appear when systematically changing variables (such as a single processing condition during moulding, or ageing temperature or time) with a particular set of mouldings. Except for the observation that increasing ageing or annealing temperature and/or time causes the internal stress parameter, σ_i, to become smaller in magnitude, no general rules have emerged.

5. CONCLUSIONS

The layer removal technique is a valuable method for measuring the residual stress levels in moulded polymers, but has been applied only to mouldings in the form of flat bars or plates, or pipes. Most moulding operations produce compressive residual stresses at the surface. A common mode of failure is by fracture initiated at a surface flaw and the presence of compressive stresses near the surface will inhibit this process and can therefore be beneficial. It is thus important to know whether the residual stress distribution is optimal. Extruded pipes produced without internal cooling have been shown to have tensile stresses near the inner wall and it has been shown that the provision of forced cooling inside the tube is a suitable remedy. On the other hand, some mouldings are found to fail in tension from crazes which develop in the interior, and adjustment of the residual stress distribution to reduce the level of tensile stress here would seem advisable.

The gradient annealing experiments have revealed that residual tensile stresses may be fairly easily introduced at the surface of mouldings made from a variety of polymers, including glassy, semi-crystalline and short glass-fibre-filled grades, and indicate that considerable loss of resistance to fracture may develop in addition to the problem of warping.

Although it is recognised that distortion of mouldings is a consequence of imbalanced stresses and does not necessarily reflect stress levels, large residual stresses are more likely to lead to distortion than small ones for if the article is located in a temperature gradient the high stresses will relax by an amount greater than with small stresses, and the resulting imbalance will be greater.

A knowledge of the residual stress levels in a moulding is therefore an important factor in the assessment of long-term serviceability, and the layer removal procedure offers the best method for obtaining the required information. The use of biaxial analysis is often required and it is recommended that more attention should be paid to this than has been in the past.

The measurement of an internal stress parameter, σ_i, from stress relaxation data by the method introduced by Kubát and Rigdahl is time-consuming and does not seem to have any direct application. The studies of stress relaxation have been of value in indicating the magnitude of the equilibrium stress, for this is really what is observed when residual stresses are measured in a moulding which has had time to reach the equilibrium state. In this case the stress is different at each location and it is difficult to predict the exact course of relaxation leading to the final distribution. It is important to note that this limit does exist, however, for it implies that significant levels of residual stress will remain after annealing unless the temperature reaches the glass transition temperature, a result confirmed by experiments.

ACKNOWLEDGEMENTS

Several phases of the work described here have received financial assistance from the Science and Engineering Research Council.

REFERENCES

1. Haworth, B., Hindle, C. S., Sandilands, G. J. and White, J. R. (1982). *Plast. Rubb. Proc. Applics.*, **2,** 59.
2. Kubát, J. and Rigdahl, M. (1975). *Int. J. Polymeric Mater.*, **3,** 287.
3. Kubát, J., Rigdahl, M. and Seldén, R. (1976). *J. Appl. Polym. Sci.*, **20,** 2799.
4. Kubát, J., Seldén, R. and Rigahl, M. (1978). *J. Appl. Polym. Sci.*, **22,** 1715.
5. Fotheringham, D. G. and Cherry, B. W. (1978). *J. Mater. Sci.*, **13,** 951.

6. Andrews, E. H. (1978). *Br. Polym. J.*, **10**, 39.
7. Knappe, W. (1961). *Kunststoffe*, **51**, 562. (English translation: *Kunststoffe: German Plastics*, **51**, 56.)
8. Adams, L. H. and Williamson, E. D. (1920). *J. Franklin Inst.*, **190**, 597 and 835.
9. Aggarwala, B. D. and Saibel, E. (1961). *Phys. Chem. Glasses*, **2**, 137.
10. Lee, E. M., Rogers, T. G. and Woo, T. C. (1965). *J. Am. Ceram. Soc.*, **48**, 480.
11. Struik, L. C. E. (1978). *Polym. Eng. Sci.*, **18**, 799.
12. Isayev, A. I., Hieber, C. A. and Crouthamel, D. L. (1981). *SPE 39th Antec*, Boston, 110.
13. Mills, N. J. (1982). *J. Mater. Sci.*, **17**, 558.
14. Williams, J. G., Gray, A. and Hodgkinson, J. M. (1981). *Polym. Eng. Sci.*, **21**, 822.
15. Krausz, A. S. and Eyring, H. (1975). *Deformation Kinetics*, New York, Wiley-Interscience.
16. Kocks, U. F., Argon, A. S. and Ashbee, M. F. (1975). *Progress in Materials Science*, Vol. 19, Chalmers, B., Christian, J. W. and Massalski, T. B. (eds), Oxford, Pergamon Press.
17. Ward, I. M. (1971). *Mechanical Properties of Solid Polymers*, Chap. 7, New York, Wiley-Interscience.
18. White, J. R. (1980). *Mater. Sci. Eng.*, **45**, 35.
19. White, J. R. (1981). *Rheol. Acta*, **20**, 23.
20. White, J. R. (1981). *J. Mater. Sci.*, **16**, 3249.
21. Ito, K. (1977). *Japan Plastics Age*, **15**, 36.
22. Broutman, L. J. and Krishnakumar, S. M. (1974). *Polym. Eng. Sci.*, **13**, 249.
23. Saffell, J. R. and Windle, A. H. (1980). *J. Appl. Polym. Sci.*, **25**, 1118.
24. Qayyum, M. M. and White, J. R. (1982). *Polymer*, **23**, 129.
25. Qayyum, M. M. and White, J. R. (1983). *J. Appl. Polym. Sci.*, **28**, 2033.
26. Read, B., Duncan, J. C. and Meyer, D. E. (1984). *Polymer Testing*, this issue, pp. 143–64.
27. Barrett, C. S. and Predecki, P. (1976). *Polym. Eng. Sci.*, **16**, 602.
28. Predecki, P. and Barrett, C. S. (1979). *J. Composite Mater.*, **13**, 61.
29. White, J. R. (1977). *Proc EMAG 77. IoP Conf. Sers.*, **36**, 411.
30. Kent, R. J., Puttick, K. E. and Rider, J. G. (1981). *Plast. Rubb. Proc. Applics.*, **1**, 55.
31. Kent, R. J., Puttick, K. E. and Rider, J. G. (1981). *Plast. Rubb. Proc. Applics*, **1**, 111.
32. Treuting, R. G. and Read, W. T., Jr. (1951). *J. Appl. Phys.*, **22**, 130.
33. Russell, D. P. and Beaumont, P. W. R. (1980). *J. Mater. Sci.*, **15**, 208.
34. Coxon, L. D. and White, J. R. (1980). *Polym. Eng. Sci.*, **20**, 230.
35. Coxon, L. D. and White, J. R. (1979). *J. Mater. Sci.*, **14**, 1114.
36. Hindle, C. S., (1980). Unpublished results.
37. Thomas, K., Dawson, D., Greenwood, W. J., White, J. R., Hindle, C. S. and Thompson, M. (1982). Reported at 5th Int. Conf. on Deformation Yield and Fracture of Polymers, Cambridge, April, 1982.

38. Hindle, C. S., White, J. R., Dawson, D., Greenwood, W. J. and Thomas, K. (1981). *39th Antec*, Boston, 783.
39. Thomas, K. and Greenwood, W. J. (1980–2). Unpublished results.
40. Li, J. C. M. (1967). *Canad. J. Phys.*, **45**, 493.
41. Haworth, B. and White, J. R. (1981). *J. Mater. Sci.*, **16**, 3263.
42. Kubát, J. and Rigdahl, M. (1975). *Mater. Sci. Eng.*, **21**, 63.
43. Clark, E. S. and Garber, C. A. (1971). *Int. J. Polymeric Mater.*, **1**, 35.
44. Clark, E. S. (1973). *Appl. Polym. Symp.*, **20**, 325.
45. Clark, E. S. (1974). *Appl. Polym. Symp.*, **24**, 45.
46. Fitchmun, D. R. and Mencik, Z. (1973). *Polym. Sci. Polym. Phys. Ed.*, **11**, 951.
47. Mencik, Z. and Fitchmun, D. R. (1973). *Polym. Sci. Polym. Phys. Ed.*, **11**, 973.
48. Bowman, J., Harris, N. and Bevis, M. (1975). *J. Mater. Sci.*, **10**, 63.
49. Bowman, J. and Bevis, M. (1976). *Plast. Rubb. Maters. Applic.*, **1**, 177.
50. Tan, V. and Kamal, M. R. (1978). *J. Appl. Polym. Sci.*, **22**, 2341.
51. Fujiyama, M. (1975). *Kobunshi Ronbunshu*, **32**, 411. (English edition, **4**, 534.)
52. Fujiyama, M. and Kimura, S. (1975). *Kobunshi Ronbunshu*, **32**, 581. (English edition, **4**, 764.)
53. Fujiyama, M. and Kimura, S. (1975). *Kobunshi Ronbunshu*, **32**, 591. (English edition, **4**, 777).
54. Fujiyama, M., Awaya, H. and Kimura, S. (1977). *J. Appl. Polym. Sci.*, **21**, 3291.
55. Fujiyama, M. and Kimura, S. (1978). *J. Appl. Polym. Sci.*, **22**, 1225.
56. Fujiyama, M. and Azuma, K. (1979). *J. Appl. Polym. Sci.*, **23**, 2807.
57. Hobbs, S. Y. and Pratt, C. F. (1975). *J. Appl. Polym. Sci.*, **19**, 1701.
58. Sandilands, G. J. and White, J. R. (1980). *Polymer*, **21**, 338.
59. Haworth, B., Sandilands, G. J. and White, J. R. (1980). *Plast. Rubb. Int.*, **5**, 109.
60. Thompson, M. and White, J. R. (1984). *Polym. Eng. Sci.*, (in press).
61. Qayyum, M. M. and White, J. R. To be published.
62. Srivastava, A. K. (1983). MSc Thesis, University of Newcastle upon Tyne.
63. Sandilands, G. J. (1983). PhD thesis, University of Newcastle upon Tyne.

Polymer Testing **4** (1984) 193–194

Component Shape and Dimensional Change Measured by Holography and Moiré Shadow Fringe Methods

A. E. Ennos and K. Thomas

National Physical Laboratory, Teddington, Middlesex TW11 0LW, UK

SUMMARY

Plastics components must commonly be manufactured to meet required dimensional tolerances. Normally inspection is carried out by mechanical measurement, e.g. in checking the roundness of a cap. With more complicated shapes, measurement becomes increasingly time-consuming and expensive. Optical methods provide a means of measurement that is non-contacting and gives an overall assessment of shape more easily. For example, in a recent development shape in three dimensions is measured by the so-called 'reflex metrograph', where a pointer is moved by hand over an image of the object surfaces and an attached pen traces out the contour followed with an accuracy of some 0·1 mm. In another technique, known as optical contouring, shape is measured by projecting a grating pattern on to the object, and recording the distorted line pattern; it is anticipated that development of computer analysis of these patterns will give an accuracy of 0·25 mm. Whilst the metrograph gives an absolute measurement of shape, the projected grating technique is really a differential method comparing an observed contour pattern with that for a flat object. Differential methods can be quicker and more accurate, and the use of two more of these was discussed.

A common requirement is to check the flatness of part of a moulding. A simple way of obtaining an overall picture of out-of-flatness is to use the Moiré shadow fringe method. This is done by illuminating the surface obliquely through a glass plate with a coarse

line grating, and observing the pattern of light scattered back through the grating. Fringes are obtained which are height contours of the surface being measured—in fact, the surface is being compared with the flatness of the grating surface. Out-of-plane distortions can be measured in the range 0·1–1 mm which is of practical importance. The method is being used to determine the effect of varied moulding conditions, and can be employed to decide optimum machine settings.

A plastics article will normally shrink and change shape in the first few minutes of cooling after moulding. It will often then contain internal stresses which can relax out and cause further change of shape, with the greater change occurring in the first few hours or days. Shadow Moiré photographs recorded at intervals will reveal the larger distortions; to check for slower changes a more sensitive holographic method is needed. Essentially holography measures the difference between two successive images of the object. The method has previously required carefully controlled laboratory conditions, but a new and simpler technique has now been developed. The object is illuminated with laser light through a photographic plate held near to its surface, and a holographic image is formed in the emulsion by interference between incident light and light reflected back from the surface. If the plate is then developed and replaced exactly in position, changes in object shape cause fringe patterns to appear. The method can detect changes of 0·1 μm, and so a very sensitive check of stability is possible.

The holographic methods can further be used to observe small distortions of a component introduced by mechanical loading (e.g. internal pressure applied to a pipe, or vibration of a sheet) or by heating. The aim is to detect any anomalies in the distortion pattern produced by imperfections arising during manufacture; or otherwise to compare distortions for different designs of a component.

Polymer Testing **4** (1984) 195–209

Electrical Resistivity Measurements of Polymer Materials

A. R. Blythe

BICC Research and Engineering Ltd, Wood Lane, London W12 7DX, UK

SUMMARY

Basic methods of measuring volume and surface resistivities are reviewed in relation to the wide range of values encountered in plastics and rubbers, including antistatic grades, conducting composites and intrinsically conductive polymers. The application of conventional electrode configurations of 2, 3 and 4 terminal types is considered first. Then the performance and reliability of various probe techniques are analysed. Finally, the observation of charge decay rates is discussed as a means of determining resistivities, and its limitations are defined.

1. INTRODUCTION

Polymers have electrical resistivities that are characteristically very high, and in most electrical applications they are essentially used as insulators.[1] There are, however, requirements for conductive grades, and base polymers are frequently modified to meet this need. Thus antistatic treatments are applied to surfaces to prevent unwanted accumulation of charge, and carbon black composites are used as flexible heating elements. Variants incorporating carbon fibres afford low resistivities too, and these are valuable for screening purposes. Polymers which possess a high degree of intrinsic electronic conductivity, usually derived from doped polyacetylene, are also under

195

Polymer Testing 0142-9418/84/$03·00 © Elsevier Applied Science Publishers Ltd, England, 1984. Printed in Northern Ireland

A. R. Blythe

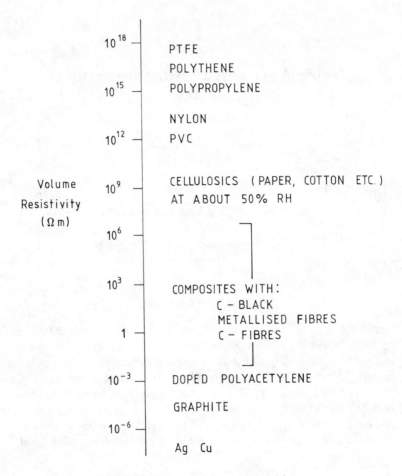

Fig. 1. Chart of volume resistivities.

development. Hence the measurement requirements for resistivity cover an exceptionally wide range as indicated in Fig. 1.

This paper, in addition to briefly reviewing the conventional methods of measuring resistivities, using specially shaped specimens and electrodes, discusses the applicability of various probe methods. The latter are attractive from the point of view of making simple measurements on moulded objects *in situ*, etc., but, unless used with great caution, they give unrealiable results. Consideration is also given to observations of the rate of decay of charge in a material as a way of determining resistivity. Although this principle is adopted in certain standard test methods, it is difficult to obtain other than comparative values by this means.

2. DEFINITIONS AND UNITS

The simple definition of volume or bulk resistivity ρ (Ωm) of an isotropic material is the resistance, as determined in accord with Ohm's law, between opposite faces of a unit cube. Whence the resistance R between opposite ends of a block of uniform cross-sectional area A and length l is given by:

$$R = \frac{\rho l}{A} \tag{1}$$

Alternatively, volume resistivity may be expressed by means of a generalised form of Ohm's law:

$$E = \rho J, \tag{2}$$

where E and J represent the electric field and current density, respectively, at any point in the material. More generally still, for an anisotropic material, resistivity is a second rank tensor that relates the two vector quantities:

$$\mathbf{E}_i = \rho_{ik} \mathbf{J}_k \tag{3}$$

Being symmetric, the resistivity tensor has three principal values when referred to its principal axes. A common case of anisotropy occurs with oriented conducting fibre composites where the resistivity component in the fibre direction (taken along the z-axis), is lower than those in the perpendicular directions, i.e.

$$\rho_{zz} \ll \rho_{xx}, \rho_{yy}$$

Volume conductivity is the reciprocal of resistivity, $1/\rho$ ($\Omega^{-1}\,\mathrm{m}^{-1}$). (It should be noted that the unit of conductance, the reciprocal ohm (Ω^{-1}), is now often called the siemen (S).)

Where current flow is confined to a surface, it is convenient to define an analogous surface resistivity σ (Ω) as the resistance between opposite edges of a square. The resistance across a square is independent of the size of the square, so that the unit of surface resistivity is properly the ohm, occasionally written rather superfluously as ohm per square. A conducting surface must in reality be a layer with a finite thickness t, and we have only an *effective* surface resistivity, which is related to the true volume resistivity of the layer by:

$$\sigma = \frac{\rho}{t} \qquad (4)$$

In some contexts it is convenient to divide materials into three categories: conductors, antistatic materials and insulators. There is no strict definition of these terms with respect to resistivity, although as a guide we can take the ranges to be $<10^3$, 10^3-10^9 and $>10^9$ Ωm, respectively. The concept of an antistatic material is that it can be used to make components that are low enough in resistance to allow rapid dissipation of static charges but high enough to provide safe insulation from mains electricity. Performance is specified[2] in more detail for certain applications.

3. CONVENTIONAL MEASUREMENT OF RESISTIVITY

In the laboratory, resistivities are most usually measured with specially shaped specimens and electrode configurations that are chosen in such a way that uniform electric fields (or field gradients) are generated in the material. Resistivities are then easily calculated from the observed currents and voltages between the electrodes.

3.1. Low volume resistivity

The simplest arrangement of all consists of a rectangular or cylindrical block with two electrodes applied, one at either end, as shown in Fig. 2(a). The resistance is taken from the ratio of the applied voltage to the series current. The main problem in accurate measurement is one of the uncertain contact resistances, between the electrodes and the specimen, that are included. Contact resistance may be reduced by replacing simple pressure contacts to metal plates and foils by silver paint, colloidal graphite (water or alcohol DAG) or vacuum-evaporated metal, but it is much better to use the four-terminal method, as shown in Fig. 2(b). A current density J is established in the central region (cross-sectional area A) by passing a known current I between the outer electrodes, and the electric field E is

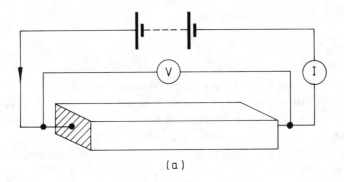

(a)

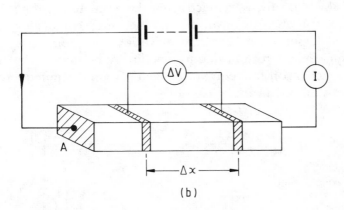

Δx

(b)

Fig. 2. Measurement of low volume resistivity: (a) two-terminal method; (b) four-terminal method.

determined by measuring the potential drop Δv across the two inner electrodes (separation Δx). The resistivity is given by:

$$\rho = \frac{E}{J} = \frac{\Delta V / \Delta x}{I / A} \tag{5}$$

Any effect of contact resistance is thereby avoided, provided that the contact resistances of the voltage electrodes are much smaller than the input resistance of the voltmeter. In practice, the method is generally restricted to resistivities below about 10^6 Ωm, otherwise currents become too small to measure accurately and voltmeter resistances become significant.

A. R. Blythe

3.2. High volume resistivity

In order to facilitate current measurement on high-resistivity materials, thin disc specimens are used (see Fig. 3), whereby the overall resistance between the current electrodes is much reduced. The main problem becomes one of leakage current from the high-voltage source to the ammeter via routes other than the intended one *through* the specimen; the surface of the specimen itself is liable to provide a low resistance path through the accumulation of dirt and moisture on it. For this reason, a three-terminal method is used, where the extra *guard* electrode is applied around the low-voltage electrode and its connection to the ammeter, so as to intercept and divert any leakage currents which are then excluded from the ammeter reading. Where very high resistivities are involved and the direct conduction current is very small, one must be aware of various transient components, e.g. a displacement current due to dipolar orientation. If the current does not reach a constant equilibrium value, then a genuine resistivity cannot be obtained in this way.

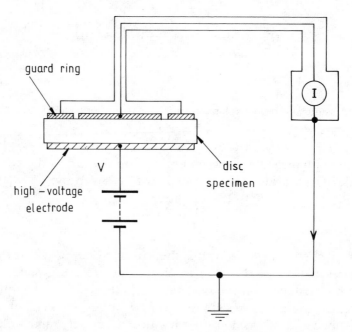

Fig. 3. Measurement of high volume resistivity: three-terminal method.

3.3. Surface resistivity

Concentric ring electrodes are the easiest to use for measurement of surface resistivity. The resistance R between them is the sum of the resistances of the elemental annuli (mean radius r) in series (Fig. 4):

$$R = \int_{r_1}^{r_2} \frac{\sigma}{2\pi r} \, dr \qquad (6)$$

where r_1 and r_2 are the radii of the inner and outer electrodes, respectively. Hence:

$$\sigma = 2\pi R \ln \left(\frac{r_1}{r_2}\right) \qquad (7)$$

This electrode geometry completely defines the area of measurement. If the surface is flat and not too hard, sufficiently good contact can be made by pressure against knife-edged metal electrodes. Otherwise it may be necessary to use conductive rubber electrodes or to paint the electrodes onto the surface. Where contact resistance remains a problem, a four-terminal method may be applied to a rectangular strip. It should be mentioned that surface conduction produced by most antistatic agents is ionic in nature and very sensitive to ambient

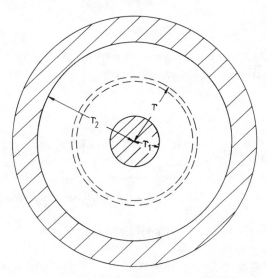

Fig. 4.   Measurement of surface resistivity.

humidity. Therefore control of relative humidity becomes an important part of surface-resistivity measurements.

4. RESISTIVITY PROBE TECHNIQUES

Some idea of the resistivity of a material may be obtained by simply probing the surface of a sheet or block with a pair of metal points connected to an ohmmeter. The result is rather difficult to interpret quantitatively, however, because the electric field patterns are complex, with large gradients near the electrode tips. As long as Ohm's law is obeyed, the local current density is still linearly related to the electric field at any point, but the current density and field vary in direction and magnitude from place to place. There is a strict analogy between the field equations of current flow with the equations of electrostatic fields, since they both comply with Laplace's equation:

$$\nabla^2 V = 0 \qquad (8)$$

where V is the electric potential at any point in the field. Thus electrodes which inject or drain current from the material are equivalent to positive and negative charges. For example, the potential at a distance r from a point electrode supplying a current I inside a conductor is given by:

$$V = \frac{I\rho}{4\pi r}, \qquad (9)$$

which may be compared with that due to a point charge q in a dielectric medium of relative permittivity ε:

$$V = \frac{q}{4\pi\varepsilon_0\varepsilon r} \qquad (10)$$

where ε_0 is the permittivity of free space. Since analytical solutions have been obtained for many charge distributions, we have a convenient fund of data for solving current-flow problems. This principle is applied below to various point probe measurements.

4.1. Two-point volume-resistivity measurement

The field equations have been solved analytically for charged prolate and oblate spheroids,[3] and this enables expressions to be devised for

TABLE 1

Resistance Between Pairs of Simple Electrodes

ELECTRODE TYPE	RESISTANCE FORMULA (Ω)
Hemispheres: $d \gg r$	$R = \dfrac{\rho}{\pi r}$
Discs: $d \gg r$	$R = \dfrac{\rho}{2r}$
Needles: $d \gg l \gg r$	$R = \dfrac{\rho}{\pi l} \ln \dfrac{2l}{r}$

the resistance between pairs of electrodes of several simple shapes. Table 1 summarises the results for various forms of electrode contacting the surface of a semi-infinite block of material having an isotropic volume resistivity. Considering the results for hemispherical electrodes—these roughly correspond to blunt metal electrodes pressed into the surface—it can be seen that the theoretical resistance is *independent of the electrode separation* (always provided that $d \gg r$, i.e. we are dealing with fine electrodes). The physical explanation is that the major part of the voltage drop occurs in the immediate vicinities of the electrode tips: we may think of the resistance as the sum of a 'spreading' resistance from the source electrode, and a 'convergence' resistance to the drain electrode.

In this treatment we have totally ignored contact resistance. We expect, and indeed find, great variability in resistance measurements made in this way due to the difficulty in making good reproducible ohmic contact with electrode tips. Furthermore, the theory shows that the method only really measures the resistivity of the material in the small region around the electrode tips, where the voltage drop occurs. When pointed or rounded electrode tips are pressed onto a pliable material the area of contact is difficult to control with precision and, in addition, the material becomes deformed by the pressure from the electrodes in just that region to which the measurement is most sensitive. Any surface skin which is less conductive than the main bulk—a common feature of some composite materials—will greatly increase the apparent resistivity too. Another spurious effect that is liable to occur in the all-important region near an electrode tip is sample overheating due to the high current density there.

For the above reasons a two-point-probe measurement of resistivity is most unreliable and cannot be recommended. If such a method is, however, especially required for some reason, it would be better to use disc-shaped electrode tips, lightly pressed onto the surface, because the material distortion would be reduced and the area of contact more precisely defined than with pointed tips.

4.2. Two-point surface resistivity measurement

In order to calculate the resistance between two electrodes contacting a conductive surface, solutions of the two-dimensional Laplace equation[4] are required. It may then be shown that the resistance between two circular electrodes of radius r at distance d apart on an infinite sheet of uniform surface resistivity is given by:

$$R = \frac{\sigma}{\pi} \cosh^{-1}\left(\frac{d}{2r}\right) \tag{11}$$

Unlike the volume-resistivity case, the resistance is not independent of electrode separation, even when the electrode separation is much larger than the electrode size.

Figure 5 illustrates an experimental check on this behaviour using Teledeltos paper, a type of recording paper that is conductive on one side by virtue of a graphitic coating. (The mean surface resistivity as determined by four-terminal measurements on strips cut from the paper was $4.5 \pm 7\%$ kΩ with a $20 \pm 5\%$ higher value in the transverse direction than in the longitudinal direction with respect to the original roll.) Two circular electrodes 2 mm in diameter were painted 50 mm apart on a large sheet, and the resistance between them measured with an AVO meter. The electrodes were then increased in size, without shifting their centres, and the resistance re-measured. In this way the effect of variable resistivity of the paper was avoided as

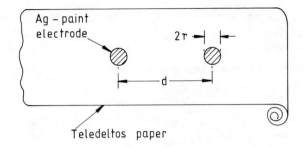

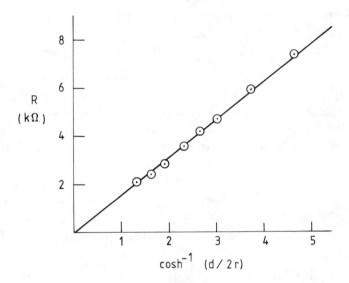

Fig. 5. Two-point-probe resistance on Teledeltos paper.

much as possible, whilst the effect of changing the electrode separation-to-diameter ratio was studied. The linearity of the graph confirms the theoretical relationship. The mean surface resistivity calculated from the slope is $4.9 \text{ k}\Omega$, which is satisfactory when the variability of the paper itself is taken into account.

The method is again sensitive to contact. Use of sharp points on soft materials is therefore unsuitable because the line of contact will depend on how far the points are pushed through the surface. Small ring electrodes gently pressed onto the surface give better control of this. Generally there is no advantage over concentric ring electrodes, except where the surface is sharply curved and two small electrodes can make better contact.

4.3. Four-point resistivity measurement

Figure 6 shows the simplest form of four-point probe measurement of volume resistivity. A row of pointed electrodes, equi-spaced distance d apart, rest on the plane surface of a semi-infinite conductor. A known current I is injected at electrode 1 and collected at electrode 4, whilst the potential difference ΔV between electrodes 2 and 3 is measured. For this arrangement Valdes[5] derived the following equation:

$$\rho = 2\pi d \frac{\Delta V}{I} \tag{12}$$

The result is independent of electrode contact area so long as the radius of contact area $r \ll d$. The measurement of volume resistivity by this method is also independent of contact resistance, provided that ΔV is measured with a sufficiently high resistance meter that ΔV is

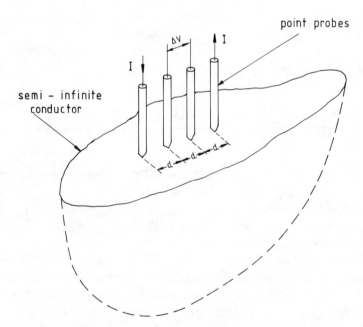

Fig. 6. Diagram of four-point resistivity probe.

not significantly affected by the meter. Consequently, the method is very reliable and is used extensively in the solid-state-electronics industry.

When the electrodes are applied to a specimen with finite boundaries the apparent resistivity ρ_a that is measured is higher than the true resistivity, and this is usually expressed in terms of a 'correction divisor' (CD):

$$\rho = \frac{\rho_a}{CD} \tag{13}$$

Uhlir[6] has derived expressions for this parameter by the method of images and has tabulated them for various geometries. For a very thin conductive sheet, i.e. thickness $t \ll d$, the expression for the CD reduces to:

$$CD = 2\frac{d}{t}\ln 2 \tag{14}$$

Since the effective surface resistivity is ρ/t (see eqn. (4)), we obtain:

$$\sigma = \frac{\pi}{\ln 2}\frac{\Delta V}{I} \tag{15}$$

Hence, when the four-point probe is used to measure a surface resistivity, the result is independent of electrode spacing.

A more general treatment of four-point probe measurements, which includes alternative electrode arrangements and takes into account anisotropic resistivities, has been presented by van der Pauw.[7]

5. CHARGE DECAY METHODS

Observation of the rate of decay of charges by leakage to earth has formed the basis for other methods of determining high resistivities, especially of antistatic films or sheets. The specimen is first charged to a high potential by one of several methods: (1) exposing the surface to a corona source of air ions; (2) rubbing with another material to produce charge transfer; or (3) connecting the edges to a high-voltage source. The charge on the material is then monitored as a function of time by means of an electric field meter facing the surface.

Quantitative interpretation of the rate of charge dissipation by conduction over a surface is complex, for it depends on the product of the system's effective resistance and capacitance to earth. These two quantities are separately distributed over the surface and related to the particular geometrical arrangement that is employed. Henry *et al.*[8] have carried out a detailed theoretical analysis which shows that, in general, the decay is not strictly exponential. Their calculation of the half-life of charge on the basis of a simple model serves, however, to indicate the typical relationship with resistivity. They consider a strip of film bridging, at right angles, two conducting planes; initially, the charge is deposited on the mid-line of the strip and the conductors are connected to earth. The half-life $t_{1/2}$ of the initial charge is then given approximately by:

$$t_{1/2} = 3 \cdot 9 \times 10^{-12} (\varepsilon + 1) \sigma a \tag{16}$$

where σ, ε are the surface resistivity and relative permittivity of the film, and a is the width of the strip. They found good experimental agreement for a film held between concentric rings, taking a to be the annular gap. From this work[8] it is clear that measurements of charge decay rate on a particular apparatus cannot easily be used to determine resistivity in any absolute way, but that the apparatus can be calibrated by means of a material of known surface resistivity and used to compare the values for a set of similar specimens. It can be very useful because the technique is quite simple to apply. Consequently the principle has been adopted in a number of standard test procedures[9] and in several types of commercial instrument.

6.   CONCLUSIONS

From the foregoing considerations we may draw the following general conclusions about resistivity measurements:

1.   For low-resistivity materials, contact resistance is of major importance so that measurement reliability generally demands a four-terminal method. The effect of contact resistance is exaggerated even more with point probes. Hence two-point probes tend to give very unrealiable results, although again a four-point version solves the problem and may be used with confidence.

2. For high-resistivity materials, contact resistance is of only minor significance and current leakage, especially over surfaces, becomes the more serious problem. For this reason, three-terminal methods using a guard electrode are commonly used. Point probe methods are usually impractical on account of the small currents they incur, requiring insruments that combine high sensitivity with high input impedance.
3. Deduction of absolute surface resistivity values from observations of charge decay rates is very complex. Nevertheless this kind of observation affords a convenient technique for comparative studies, especially for antistatic films and sheets.

REFERENCES

1. Blythe, A. R. (1979). *Electrical Properties of Polymers*, Cambridge, Cambridge University Press.
2. BS 2050 (1978). *Electrical Resistance of Conducting and Antistatic Products made from Flexible Polymeric Material.*
3. Moon, P. and Spencer, D. E. (1961). *Field Theory for Engineers*, New York, van Nostrand Co.
4. Peek, F. W. (1929). *Dielectric Phenomena in High Voltage Engineering*, Chapter 2, New York, McGraw-Hill Book Co.
5. Valdes, L. B. (1954). *Proc. Inst. Radio Engrs.*, **42,** 420.
6. Uhlir, A. (1955). *Bell Systems Tech. J.*, 105.
7. van der Pauw, L. J. (1961). *Phillips Res. Repts.*, **16,** 187.
8. Henry, P. S. H., Livesey, R. G. and Wood, A. M. (1967). *J. Textile Inst.*, **58,** 55.
9. BS 2782:Part 2:Method 250A (1976). *Antistatic Behaviour of Film. Charge Decay Method.*

Polymer Testing **4** (1984) 211–223

Measurement of Rubber Properties for Design

M. J. Gregory

The Malaysian Rubber Producers' Research Association, Tun Abdul Razak
Laboratory, Brickendonbury, Hertfordshire SG13 8NL, UK

SUMMARY

*The rubber properties required for the design of engineering compo-
nents are primarily concerned with stiffness, although factors such as
strength and ageing resistance must be taken into account in materials
selection. Calculation of static behaviour requires a shear modulus at
the appropriate strain. In order to allow for the strain dependence of
the shear moduli of filled rubbers, it is preferable to measure moduli
over a range of strains rather than to rely on a single point measurement.*

*For the purposes of vibration isolation, the stiffness of the rubber is a
function not only of frequency but also of the amplitude of the
vibration. To allow for this, the shear moduli of the rubber at low
strains, of the order of 1%, are required, over a range of frequencies.*

1. INTRODUCTION

The measurement of properties for engineering design purposes must
yield information which is both functional and quantitative, as opposed to indicative or comparative. Conventional rubber tests, such
as 'modulus', tensile strength and resilience, tend to be of the latter
type, and of no immediate value to a design engineer.

Polymer Testing 0142-9418/84/$03·00 © Elsevier Applied Science Publishers Ltd,
England, 1984. Printed in Northern Ireland

The properties of rubbers are often not well understood by engineers more used to using steel or concrete where deflections are small and where component behaviour can be predicted from a small number of material constants, such as Young's modulus and bulk modulus. Rubber components for engineering use can be designed using a fairly small number of design parameters, but these require somewhat different definitions to those commonly encountered in metals or plastics.

Rubber components perform a variety of tasks in both civil and mechanical engineering, ranging from essentially static applications, such as bridge bearings, to vibration or shock isolation. In these spring applications the stiffness characteristics of the component are the major factors determining suitability for a given task, although other considerations such as strength or ageing resistance must be taken into account at a later stage.

In the initial stages of design, the rubber properties required are the static and dynamic moduli, including loss angle (which is determined by the ratio of the out-of-phase to the in-phase moduli). This paper will concentrate on the identification and measurement of the rubber moduli relevant to component design.

2. CHARACTERISATION OF STATIC STIFFNESS

The static stiffness or compliance of a rubber spring is one property of considerable interest to the designer. Numerous solutions[1] to load–deflection relationships have been given, usually involving the Young's modulus of the rubber. A typical example is compression of a rubber disc bonded between steel plates. A theoretical solution[2] for the load–deflection relationship is that:

$$F = E_0 A (1 + 2S^2) x/t \tag{1}$$

where S is a shape factor, defined in the case of a disc as $D/4t$ (D being the diameter and t the thickness), E_0 is the 'Young's modulus' of the rubber, A is the bonded area, and x is the deflection. In this instance, as in many others, the quantity E_0 is a property of vital importance to the designer, but is often not adequately defined.

Rubbers exhibit non-linear elasticity so that the stress/strain ratio varies markedly with the type and amplitude of the applied strain.

Elasticity theories developed for large deformations of incompressible solids predict that the stress (σ_c) exerted by a rubber in *homogeneous* compression is given by:

$$\sigma_c = G_c'(\lambda - \lambda^{-2}) \tag{2}$$

where λ is the extension ratio t/t_0 ($= 1 + e$), and G_c' is a parameter which can vary not only with the strain but also with the type of strain, i.e. tensile or compressive. At infinitesimal strains, G_c' becomes equal to $E_0/3$ and eqn. (2) reduces to Hooke's law.

Similarly in shear, these theories predict that the shear stress (σ_s) is given by:

$$\sigma_s = G_s'\gamma \tag{3}$$

where γ is the shear strain and G_s' is a parameter which, like G_c', may vary with strain.

There is no a priori reason why G_s' should equal G_c' except at infinitesimal strain, and the classical relationship that $E = 3G$ may not apply at finite strains. The question then arises as to whether one can characterise the load–deflection behaviour of rubbers by a small number of parameters determined from a single testing regime.

It is apparent from eqn. (2) that the usual definition of modulus (i.e. stress/strain) is of limited applicability to rubbers. Even if G_c' is a constant this modulus will alter with strain in both tension and compression. A more appropriate replacement for modulus would be the ratio $\sigma/(\lambda - \lambda^{-2})$, ($= H$) which removes the non-linearity associated with incompressibility of the rubber. This ratio is an effective shear modulus and, in an ideal rubber where G_c' is strain-independent, should equal the shear modulus G_s' ($= \sigma_s/\gamma$).

For unfilled rubbers, quite good correspondence can be obtained between the shear modulus and the effective shear modulus obtained from tensile or compressive deformations in this way (Fig. 1). At higher tensile strains the familiar deviations described by the Mooney relationship became apparent, but these high strains are of little relevance to rubber springs where mean strains of less than 20% are most common.

For black-filled rubbers, which are of greater practical importance, the shear moduli calculated from tension or compression are close to the measured shear moduli, if the tensile shear and compression strains are the same, but systematic differences can be seen. It is also

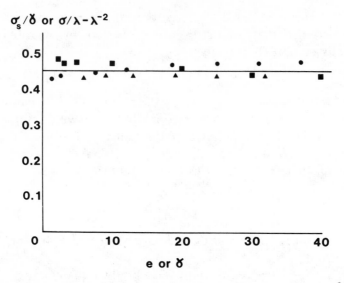

Fig. 1. Relationship of shear modulus (G) and the function $\sigma/(\lambda - \lambda^{-2})$ (or H) for an unfilled rubber in various modes of deformation. ▲, Compression; ●, shear; ■, tension.

apparent that the moduli are no longer independent of strain but show considerable variation, particularly below 10% strain (Fig. 2). A closer relationship is obtained between H and G'_s if the two moduli are compared at equal values of the strain invariant I_1, where:

$$I_1 - 3 = \lambda^2 + 2/\lambda \qquad \text{(in tension or compression)}$$

$$= \gamma^2 \qquad \text{(in shear)} \qquad (4)$$

For low to moderate strains, compressive or tensile deformations give the same value of I_1 at a given shear strain if:

$$\gamma = \surd(3)e \qquad (5)$$

The shear modulus obtained for a compressive strain e will therefore be the same as the shear modulus at a shear strain of approximately $2e$ (Fig. 3).

It follows that tensile testing of rubbers can provide useful design data if analysed in a suitable way. While conventional rubber moduli are of little value, owing to the high strains of measurement, low strain (i.e. 1–50%) tensile moduli can be used to define the static shear moduli of a rubber fairly precisely. It is necessary, however, to

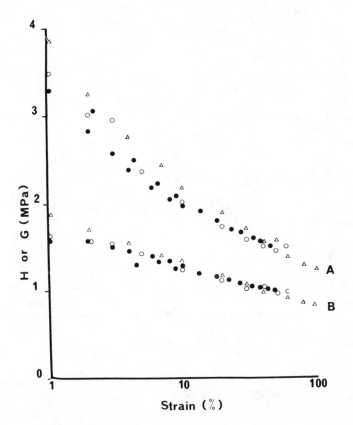

Fig. 2. Variation of H and G with strain for filled rubbers. Rubber A contains 60 pphr of an HAF carbon black, while B contains 40 pphr. $\triangle$, Shear; $\bigcirc$, tension; $\bullet$, compression.

define the strains at which measurements are made because of the variation in modulus associated with different levels and types of strain. Provided these are known, it is immaterial how the 'modulus' is expressed, as the stress/strain ratio is also a useful parameter. Use of a 'Young's modulus' of a rubber at an unspecified strain can, however, lead to significant errors in predicted stiffnesses or deflections.

The relationships given above refer to homogeneous deformations but many rubber springs are subjected to inhomogeneous deformations. The bonded disc mentioned earlier, for example, undergoes internal shear deformations which vary both across the diameter and

216 M. J. Gregory

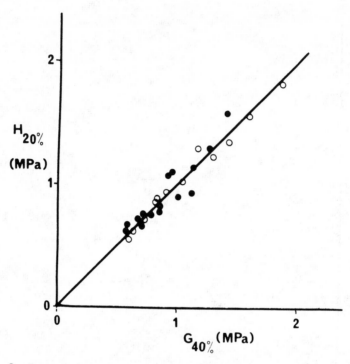

Fig. 3. Interrelation of compressive or tensile strains and shear strains from eqn. (5). The line shows the predicted relationship. ○, Tension; ●, compression.

through its thickness. It follows, therefore, that the load–deflection relationships for such a unit cannot be described using a modulus obtained for a single strain. Finite element analysis may eventually be used to predict these relationships, but a quantitative relationship between modulus and shear strain may be required. In the absence of such analysis the load–deflection relationships given for rubber components, such as eqn. (1), can only be regarded as approximate, although most rubber components are subjected to strains of such a magnitude that the change in modulus with strain is fairly small. Another complication is that although rubbers are generally regarded as incompressible significant amounts of bulk compression are likely to occur at values of S which are greater than about 5. For design of such units, the bulk modulus is required in addition to the shear modulus.

3. PROVISION OF DATA FOR STATIC STIFFNESS

In order to provide a designer with as much information as may at
present be required, MRPRA (Malaysian Rubber Producers' Research Association) publish a series of data sheets on natural rubber
vulcanisates. These provide shear moduli, bulk moduli and compression moduli of bonded units for a wide range of natural rubber
formulations. Static shear moduli from 2 to 350% shear strain are
provided at ambient temperature, enabling a designer either to select
the modulus at a suitable strain or to express the shear modulus as a
suitable function of strain. Moduli are provided for both initial
deformation and deformation after nine cycles to 100% shear strain.

In view of the uncertainties associated with inhomogeneous strain
situations, modulus data are also provided for bonded discs having
various shape factors, in compression, again for a range of strains.

The test pieces used are shown in Fig. 4. The cylindrical shear
sample was chosen for the following reasons:

1.   Ease of clamping in a rigid jig enables oscillation about zero

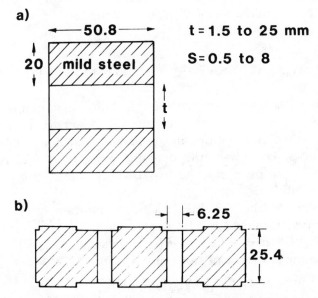

Fig. 4. Test pieces for: (a) compression, and (b) shear. All dimensions in millimetres.

strain and allows the same type of sample to be used for dynamic testing.

2. Clamping of the end units avoids the contraction perpendicular to the direction of shear which occurs with quadruple shear tests.

3. As the samples are transfer moulded, the small sample reduces the likelihood of anisotropy.

The compression test pieces are a compromise, being the largest samples which can be tested on a 10-t test machine. Even so, allowances must be made for deflection of the steel end pieces under this load.

Further information on the static shear moduli is also provided in the form of low-frequency dynamic tests (0·1 Hz) over a wide range of temperatures. As the dynamic shear modulus at low frequencies is essentially the same as the static modulus, the designer is presented with shear moduli covering a wide range of both strains and temperatures.

Static testing is carried out under computer control, and load and displacement values are stored for analysis. In this way static moduli over a range of strains can be obtained with little more effort than required for determination of a point modulus.

4. CHARACTERISATION OF DYNAMIC STIFFNESS

Static shear moduli are required for calculation of initial deflections of bearings, but most rubber bearings are used to accommodate motions of some description, in applications such as vibration or shock isolators. For these, the dynamic modulus of the rubber is required or, more precisely, the in-phase and out-of-phase components of the modulus.

For a simple linear viscoelastic solid, imposition of a sinusoidal strain cycle results in a sinusoidal stress cycle which is at an angle δ in advance of the strain. The following relationships can be used to characterise the response of the solid:

$$\text{Peak stress/peak strain} = G^* \tag{6a}$$

$$\text{Tan } \delta = G''/G' \tag{6b}$$

$$G^* = \sqrt{(G'^2 + G''^2)} \tag{6c}$$

where G^* is the complex modulus, G' is the in-phase modulus, G'' is the out-of-phase modulus and δ is the loss angle. To characterise the material, two of these parameters must be known.

In principle, the measurement of complex modulus and loss angle is simple, provided the displacement applied to the sample and the resulting load are monitored using suitable transducers. Peak loads and displacements can be monitored using voltmeters, and the phase shift measured using a phase meter. Alternatively, Fourier analysis of the output signals can be used to obtain these quantities.

Unfortunately, rubbers are frequently non-linear elastic materials, and the simple analysis above is not strictly valid. Input of a sinusoidal strain cycle can lead to a stress output in which higher harmonics play a significant role. Fortunately, these are usually of minor significance, except where high-amplitude compressive strain cycles are involved.

The use of rubber springs as antivibration mounts relies on the frequency of the input vibration being at least 1·5 times the natural frequency of the spring mass combination. The natural frequency (W_0) of the system is given by:

$$W_0 = \frac{1}{2\pi} \sqrt{\frac{K}{m}} \tag{7}$$

where K is the spring stiffness, and m the supported mass. The efficiency of isolation, however, is determined not by the ratio of the disturbing frequency to the natural frequency, but by the ratio of the disturbing frequency to an effective natural frequency corresponding to the frequency and amplitude of the input.

The effective natural frequency is given by

$$W_0' = \frac{1}{2\pi} \sqrt{\frac{K_{w,e}}{m}} \tag{8}$$

where $K_{w,e}$ is the spring stiffness at the frequency 'w' and amplitude 'e' of the disturbing cycle. The effects of frequency on shear modulus are well documented, but much less attention is paid to the effects of amplitude. In service a vibration isolator is subject to a static deformation by the load m, and a disturbing vibration is superimposed on this deformation. The stiffness towards the vibration is given by the change in force divided by the change in displacement. For an ideal non-hysteretic rubber, the stiffness in a shear deformation is given by

M. J. Gregory

differentiation, as:

$$K_w = GA/t \tag{9}$$

while for compression:

$$K = \frac{A(1+2S^2)}{t}\frac{d\sigma}{de}$$

$$= \frac{A(1+2S^2)}{t}(1+2\lambda^{-3})G \tag{10}$$

Thus the dynamic stiffness can be obtained from the slope of a load–deflection curve obtained at the required frequency.

For the vast majority of practical rubbers, these simple relationships no longer apply. Imposition of a vibration on a deformed rubber results in partial retraction, and it is the stiffness in retraction, rather than in extension, which determines the stiffness towards a vibration (Fig. 5). The retraction stiffness must be higher than the extensional stiffness, and use of the tangent modulus to predict

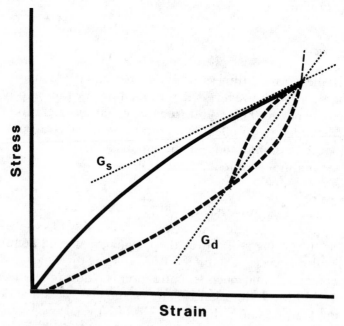

Fig. 5. Relationships between 'static' and 'dynamic' moduli for a rubber subjected to vibration about a mean strain.

dynamic stiffness can result in serious underestimates of the natural frequency with consequent lower efficiency of isolation than predicted.

From eqns. (9) and (10) it is possible to obtain an effective dynamic shear modulus (G_d) which can be a function not only of frequency but also of the amplitudes of both static and dynamic strains.

Figure 6 shows how this dynamic modulus of a heavily filled natural rubber vulcanisate is influenced by the amplitudes of both static and dynamic strains. The dynamic shear modulus is almost independent of the static strain imposed, but the dynamic strain amplitude has a marked effect on the dynamic modulus.

The results in Fig. 6 show that the dynamic shear modulus when a static strain is applied can be estimated from the conventional dynamic shear modulus measured in a symmetrical test, i.e. when cycled through zero strain. However, as the dynamic shear modulus is

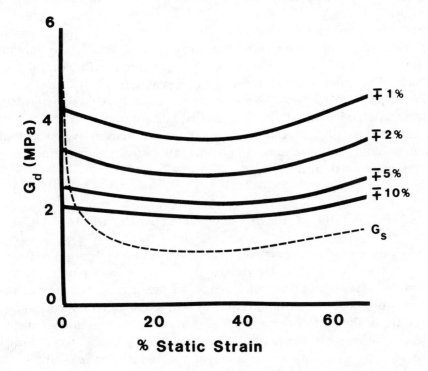

Fig. 6. Variation of 'dynamic' modulus with static strain and vibration amplitude. The broken line shows the tangent modulus from a static stress strain curve.

influenced by changes in strain, rather than by the absolute strain, the appropriate conventional shear modulus is that at a zero-to-peak amplitude corresponding to the peak-to-peak amplitude of the dynamic strain applied to the deformed rubber.

The dynamic shear modulus can also be used to predict compression stiffnesses from eqn. (10) if allowances are made for the differences between compression and shear strains. It is therefore apparent that, in addition to static shear measurements at moderate strains, dynamic shear moduli at low strains are required for prediction of dynamic stiffness. Simply measuring the slope of a force–deflection curve can lead to serious errors in prediction of dynamic stiffness of components. For the rubber used to obtain the results given in Fig. 6, the tangent modulus (i.e. $d\sigma/d\gamma$) for a shear deformation is up to four times lower than that obtained if the rubber is allowed to retract. Consequently, the natural frequency predicted would be one half of the actual value, possibly resulting in magnification of the unwanted vibration.

In order to enable a design engineer to predict the dynamic stiffness of components, MRPRA provides dynamic shear moduli over a range of strains from 1 to 100%, at frequencies from 0·1 to 15 Hz. The moduli are obtained using a symmetrical shear deformation applied to the test pieces described earlier. Although the frequency range is limited, the data is adequate for natural rubber which is known to give moduli of low-frequency sensitivity, except at sub-ambient temperatures. For other rubbers a higher frequency range may be desirable.

5.   PROVISION OF DATA ON DYNAMIC BEHAVIOUR

The other material property of great importance in vibration isolation is the loss angle. This determines the peak transmissibility of the mounted system and also, although to a lesser extent, the transmissibility well above the natural frequency. Although antivibration mountings are designed to operate well above the natural frequency of the system, vibration at the natural frequency may be encountered during the starting up or shutting down of machinery, or low-amplitude vibrations at the natural frequency may be present in the vibration spectrum.

The loss angle of a rubber varies with temperature and frequency and, if substantial amounts of fillers are used, with the amplitude of dynamic deformation. Information on the loss angles obtained with natural rubber is provided for a wide range of amplitudes, frequencies and temperatures in order that a designer can estimate the peak transmissibility in a given application.

The combination of static and dynamic moduli, including loss angle, discussed here is necessary information for the initial design of rubber springs. It will be obvious that these properties cannot be obtained from single point measurements, but that a range of measurements is required to describe the mechanical behaviour of these polymers. Provided the non-Hookean behaviour of rubbers is accepted, characterisation of the material properties introduces no major problems except for the increased number of measurements required. It is believed that the extra effort involved is a small price to pay to convince engineers that natural rubber is a well-defined engineering material.

ACKNOWLEDGEMENT

The Board of MRPRA is thanked for permission to publish this paper.

REFERENCES

1. Lindley, P. B. (1978). *Engineering Design with Natural Rubber*, MRPRA, Hertford.
2. Gent, A. N. and Lindley, P. B. (1959). *Proc. Instn. Mech. Engrs.*, **5,** 354.

Polymer Testing **4** (1984) 225–249

Determination of Non-Linear Dynamic Properties of Carbon-Filled Rubbers

G. D. Dean, J. C. Duncan and A. F. Johnson

National Physical Laboratory, Teddington, Middlesex TW11 0LW, UK

SUMMARY

A forced non-resonance test method is described for determining the dynamic mechanical properties of polymeric materials over wide ranges of strain and frequency. The use of this method for carrying out studies on carbon-filled rubbers is illustrated by results which demonstrate the variation of the dynamic shear modulus and damping factor of a tyre tread material with dynamic strain amplitude, frequency and temperature. Procedures are discussed for the analysis and presentation of such data.

Two methods are described for the determination of loss factor, and results from these are compared in order to assess the validity of phase angle measurements on non-linear materials.

Brief reference is made to dynamic testing under compressive and combined compression and shear modes of deformation. The prediction of performance under this combined loading situation from experimental data obtained in shear is demonstrated.

1. INTRODUCTION

The dynamic mechanical properties of unfilled rubbers at small dynamic strain depend upon temperature and frequency. The presence of carbon black filler introduces, in addition, a dependence upon dynamic strain amplitude, which is the reason why carbon-filled rubbers are referred to here as non-linear materials. A great deal of data is therefore needed in order to specify the behaviour of a carbon-filled rubber for a range of dynamic loading conditions. This information is required to enable the accurate prediction of component performance in various applications, for example motor vehicle tyres and antivibration mountings.

This paper describes the measurement of the dynamic mechanical properties of carbon-filled rubbers over ranges of temperature, frequency and dynamic strain amplitude. Analytical procedures are discussed which reduce the amount of testing needed to fully specify a materials behaviour and simplify the presentation of data. The measurement techniques described are applicable to the dynamic testing of a wide range of polymeric materials, whilst the analyses and results presented here concentrate on the non-linear behaviour of carbon-filled rubbers.

In the first part of this paper dynamic shear modulus and loss factor results are presented for a carbon-filled rubber used in motor vehicle tyres. A comparison between two methods of obtaining loss properties is also described, the purpose of this being to investigate the validity of phase angle measurements on non-linear materials.

The second section of this paper is concerned with the determination of the dynamic mechanical behaviour of carbon-filled rubbers in compression and combined compression and shear loading situations. Procedures, developed in earlier work, were used to relate these results to appropriate shear data, thereby resulting in a method of predicting behaviour under more complex loading conditions from shear data alone.

2. DEFINITION OF DYNAMIC MODULUS AND LOSS FACTOR

When a sinusoidal displacement is applied to a linear viscoelastic material the response is a sinusoidal stress having the same frequency f but leading the strain by a phase difference δ, called the phase angle (see Fig. 1).

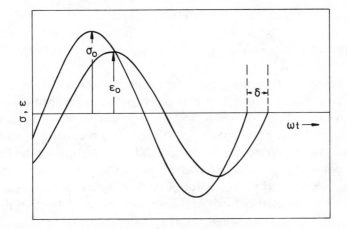

Fig. 1. Out-of-phase oscillating stress (σ) and strain (ε) for a linear viscoelastic material.

If the stress and strain cycles are taken to be represented by the real parts of

$$\sigma(t) = \sigma_0 \exp[i(\omega t + \delta)] \qquad (1)$$

and

$$\varepsilon(t) = \varepsilon_0 \exp(i\omega t) \qquad (2)$$

respectively, where $\omega = 2\pi f$, then the stress:strain ratio at any instant in time defines a complex modulus

$$M^* = \frac{\sigma_0}{\varepsilon_0} \exp(i\delta) \qquad (3)$$

This can be resolved into real (M') and imaginary (M'') components such that

$$M^* = M' + iM'' \qquad (4)$$

and

$$M' = \frac{\sigma_0}{\varepsilon_0} \cos\delta, \qquad M'' = \frac{\sigma_0}{\varepsilon_0} \sin\delta \qquad (5)$$

M' is termed the storage modulus and is proportional to the maximum energy stored per cycle whilst M'' is termed the loss modulus and is proportional to the energy dissipated per cycle.

The loss factor ($\tan\delta$ or d) is defined by the ratio

$$d = \tan\delta = M''/M' \qquad (6)$$

3. EXPERIMENTAL

3.1. Apparatus

Measurements of dynamic moduli have been carried out on dynamic mechanical apparatus designed and constructed at the National Physical Laboratory. This has been described in earlier work[1] and recent modifications have only concerned the data processing electronics. The apparatus is based upon the forced non-resonance technique, which is particularly suitable for the study of low-modulus, high-loss materials such as carbon-filled rubbers. The essential features of the dynamic mechanical apparatus are described below, whilst an illustration with the shear load stage fitted can be seen in Fig. 2.

Samples can be accommodated in a variety of load stages, such as tensile, compression, shear and combined compression and shear. An electromagnetic vibrator is driven from a signal generator (sinusoidal output) having a frequency range of 0·01–1000 Hz. The moving vibrator table is connected to one half of the load stage, and a linear

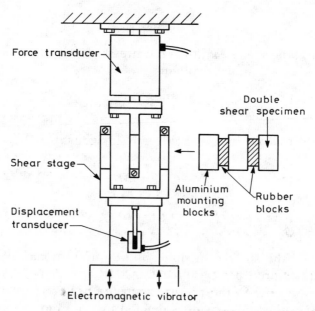

Fig. 2. Schematic of dynamic mechanical apparatus showing shear load stage and double shear specimen.

variable differential transformer (LVDT) transducer is used to measure its displacement. The other half of the load stage is bolted to a rigidly mounted force transducer. The transducers are connected to matched bridges, so that no significant extraneous phase angle is introduced, and the bridge outputs are fed to one of two processing devices. The first consists of a voltmeter and an accurate crystal-controlled timer which measures the maxima and minima of the force and displacement waveforms and the time interval between the centres (zero volts if no d.c. offset) of each waveform. The sensitivity with which phase angles can be recorded is better than $0 \cdot 1°$, but the accuracy of measurements is generally less than this due to noise on, or distortion of, the waveforms. However, errors developed in this way can usually be improved by filtering and, for unfilled rubbers, the accuracy in recording δ would usually be expected to be better than $\pm 0 \cdot 5°$ ($\pm 0 \cdot 01$ in tan δ).

The second method of signal processing involves the use of a transient recorder (Datalab DL2002). This was added to the original apparatus for the purpose of checking the validity of loss property determinations on non-linear materials by direct phase angle measurements (see next section). This equipment simultaneously digitises (using a 10-bit A/D converter) the output from the stress and strain bridges at discrete time intervals, typically recording two cycles of each waveform. Thus 4096 discrete voltage values are stored in an electronic memory for each waveform and, as this includes two cycles, a typical loop will be constructed from approximately 2000 coordinates (see Section 3.2). The resultant accuracy from this data is $\pm 1\%$ for moduli values, with a resolution of $0 \cdot 1°$ for phase angle measurements. Again the presence of noise on either waveform will lead to lower accuracies, although for measurements carried out in this work these were insignificant.

Both of the above devices are connected to a computer which calculates moduli and loss factors.

3.2. Measurements on carbon-filled rubbers

An investigation of the influence of frequency, temperature and dynamic strain amplitude on the dynamic properties of rubber is most conveniently carried out under a simple shear deformation. A suitable design for load stage and test specimen is given in Fig. 2. The complex shear modulus, G^* is, by analogy with eqn. (3)

$$G^* = \tau(t)/\gamma(t) \tag{7}$$

where $\tau(t) =$ dynamic shear stress = applied force/sample cross-sectional area, and $\gamma(t) =$ dynamic shear strain = displacement/thickness.

Depending on sample dimensions, it may be necessary to apply corrections for sample bending under simple shear deformation and for the compliance of the dynamic mechanical apparatus (see Sections 2.2 and 2.3 in Reference 1 for details).

The specimen used is a double shear arrangement (see Fig. 2) and consists of two identical samples (rectangular, square or circular cross-section) bonded to aluminium mounting blocks. A room-temperature curing epoxy resin is suitable for this purpose. The low mass and good conductivity of the aluminium blocks ensure that the temperature of the rubber samples quickly follows any changes of test temperature.

The materials studied were conditioned prior to testing by subjecting the shear sample to a higher strain than it would experience in the testing for a period of 1 min. The sample was then allowed to reach thermal equilibrium. This conditioning schedule was carried out to ensure that reproducible results were obtained since the moduli of carbon-filled materials are highly dependent on recent strain history, particularly at low strain levels. This behaviour is believed to be associated with the structural breakdown of the reinforcing carbon black at higher strain (References 2 and 3 (Section 3.3)). Thus the material's dynamic modulus will exhibit a marked strain-amplitude dependence, particularly below strain levels of 0·01.

The non-linear behaviour of these materials also gives rise to another problem, namely one of distortion of the transducer output signal. For such materials, the stress response to an applied sinusoidal strain will be a distorted sinusoid. Although a time lag still exists between each waveform, its value depends upon the nature of the distortion and varies with time throughout a cycle. There is some concern therefore, whether a measurement of phase angle will allow a valid determination of loss properties. It is for this reason that an alternative method of phase angle measurement using a transient recorder was incorporated.

In this method, calculations are based on the stress–strain loop for

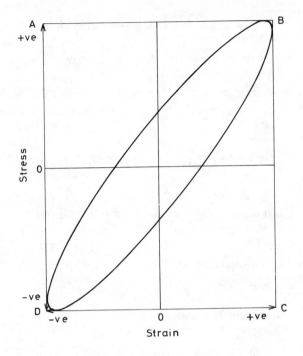

$$\sin \delta_a = \frac{4 \times (\text{area enclosed within the loop})}{\pi \times (\text{area of circumscribing rectangle ABCD})}$$

$$M' = \frac{BC}{AB} \cos \delta_a \qquad\qquad M'' = \frac{BC}{AB} \sin \delta_a$$

Fig. 3. Loop method of calculating M' and δ_a.

one cycle of deformation. Such a loop is produced by cross plotting instantaneous stress and strain values for one cycle. A computer performs this task, calculating the area of the loop and the magnitude of its circumscribing rectangle (see Fig. 3). The required quantities can then be obtained as indicated in References 3 (Section 3.4) and 4

$$\sin \delta_a = \frac{4 \times (\text{area enclosed within the loop})}{\pi \times (\text{area of circumscribing rectangle ABCD})} \qquad (8)$$

$$\frac{\tau_0}{\gamma_0} = \frac{BC}{AB} \qquad (9)$$

The ratio τ_0/γ_0 can be substituted for the ratio σ_0/ε_0 in eqn. (5) to give G' and G'', whilst the loss factor is simply $\tan \delta_a$, where δ_a is the phase angle between two undistorted waveforms which would give the same area enclosed by the stress–strain loop. An ellipse results from linear viscoelastic materials, but when the modulus is dependent on the strain level a distorted ellipse is produced (see Section 4.2).

4.   SOME ILLUSTRATIVE RESULTS

In this section, data are presented on the dynamic shear modulus G' and loss factor $\tan \delta_a$ of a carbon-filled rubber, used in motor vehicle tyre applications, and their variation with frequency, temperature and dynamic shear strain amplitude. The material was SBR, supplied by the Avon Rubber Company Ltd, with 77 phr N 339 carbon black and $\simeq 40$ phr oil.

4.1. Variation of dynamic shear modulus G'

Figures 4, 5 and 6 show the variation of G' with temperature, frequency and shear strain amplitude, respectively. This variation is quite considerable and G' can be seen to range between $1 \cdot 21 \text{ MN m}^{-2}$ ($T = 25\,°\text{C}$, $f = 0 \cdot 2$ Hz, $\gamma = 0 \cdot 10$) and $9 \cdot 36 \text{ MN m}^{-2}$ ($T = 25\,°\text{C}$, $f = 100$ Hz, $\gamma = 0 \cdot 00125$), Trends may be readily observed. G' decreases exponentially with increasing temperature, for constant frequency and dynamic shear strain amplitude. The variation with frequency also follows an exponential relationship, G' increasing with increasing frequency, for constant temperature and dynamic shear strain amplitude. Finally the marked non-linear dependence of G' upon the dynamic shear strain amplitude can be seen in Fig. 6. The scale for shear strain is split to emphasise the large variation at small strain levels. This is a typical example of the structural breakdown associated with carbon-filled rubbers.

4.2. Comparison of $\tan \delta$ measured by two methods

For purposes of comparison of loss-factor values obtained from the two methods described in Section 3.1, stress and strain waveforms for a particular applied strain were analysed by each piece of apparatus

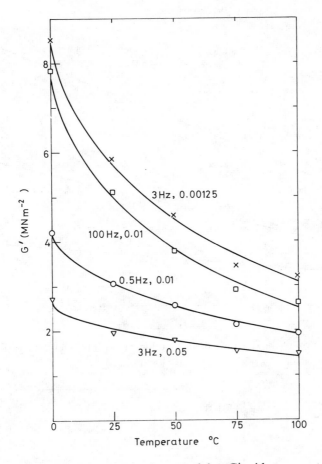

Fig. 4. Variation of the dynamic shear modulus G' with temperature at various frequencies and strain amplitudes.

consecutively so that measurements were made on identical signals. Figure 7 shows a stress–strain loop obtained from a sample undergoing a 2% dynamic shear strain amplitude where the loss factor has a maximum value, and it can be seen that there is significant deviation from an ellipse. Table 1 shows values of $\tan \delta$ obtained by both methods over a range of shear strain amplitudes at 3 and 70 Hz. Despite the distortion evident in Fig. 7, the data in Table 1 demonstrate that the difference between values of loss factor is typically 0·01, which is of similar magnitude to the error expected. This error is not systematic and it may be concluded that both methods are valid

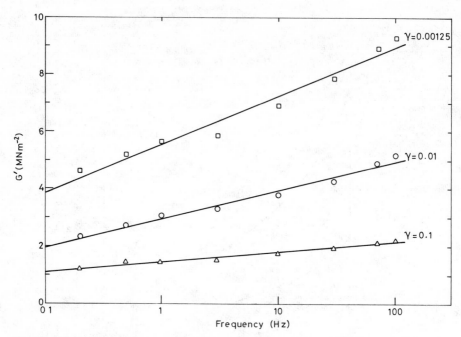

Fig. 5. Variation of the dynamic shear modulus G' with frequency at various strain amplitudes, γ.

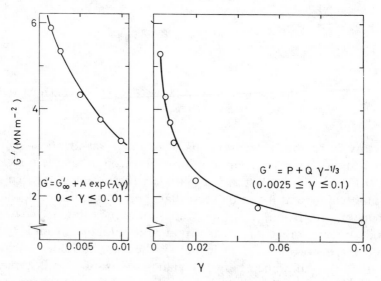

Fig. 6. Variation of the dynamic shear modulus G' with shear strain amplitude $(T = 25\,^\circ\mathrm{C}$ and $f = 3$ Hz$)$.

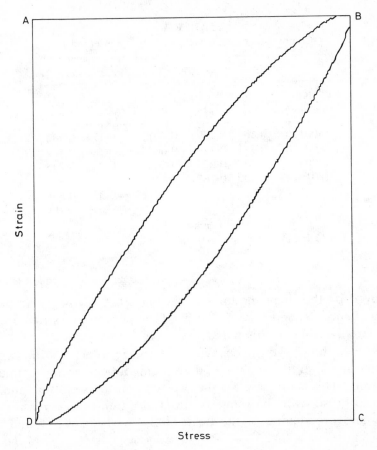

Fig. 7. Stress–strain loop for a carbon-filled rubber ($\gamma = 0\cdot02$ and $f = 3$ Hz).

for loss-factor measurement for this material. Whether both methods are satisfactory for all classes of carbon-filled rubbers would, however, need to be established by further tests.

4.3. Variation of shear loss factor (tan δ or d)

Table 1 shows that the loss factor has only a small frequency dependence, the value of d at 100 Hz being between 10 and 20% higher than that at 0·2 Hz for all values of temperature and strain. This should be contrasted with the significant frequency dependence

TABLE 1

Comparison of tan δ Measured by Two Methods

3 Hz			70 Hz		
Strain amplitude	d_{cen}[a]	d_{loop}[b]	*Strain amplitude*	d_{cen}[a]	d_{loop}[b]
0·00120	0·155	0·159	0·00124	0·174	0·182
0·00240	0·171	0·170	0·0025	0·184	0·184
0·005	0·232	0·224	0·005	0·241	0·238
0·01	0·292	0·287	0·01	0·316	0·308
			0·02	0·374	0·355

[a] d_{cen} is tan δ from zero cross-over method.
[b] d_{loop} is tan δ from stress–strain loop method.

displayed by the shear modulus G' in Fig. 5. Accordingly, trends in the frequency dependence of d at fixed temperature or strain are not considered further in this paper.

Figure 8 shows typical behaviour of the loss factor with varying dynamic shear strain amplitude, there being a peak at approximately 2% strain. This is the case for the two temperatures plotted, and was found to be so over the range 0–100 °C. A trend of decreasing loss factor and broadening of the peak at 2·0% strain was found with increasing temperature.

5. ANALYSIS OF SHEAR DATA FOR CARBON-FILLED RUBBERS

It can be appreciated that a full presentation of the temperature, frequency and strain dependence of the dynamic propeties of a carbon-filled rubber would involve a great deal of data, and extracting the required information would be a lengthy process. In this section, two methods of simplifying data presentation are suggested. The first method is an empirical curve-fitting exercise, whilst the second method enables the calculation of a complete data set from a much-reduced quantity of data.

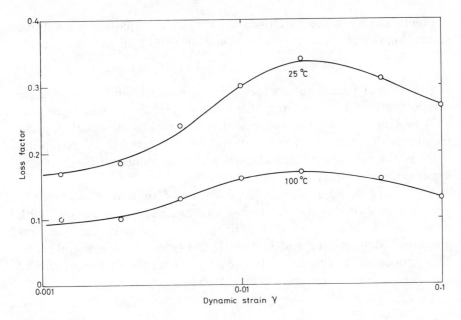

Fig. 8. Variation of loss factor with strain amplitude at two temperatures and a frequency of 3 Hz.

5.1. Curve-fitting exercises

Figures 4–6 show the variation of dynamic shear modulus (G') with temperature, frequency and shear strain amplitude, respectively. The plotted points are actual measurements, whilst the solid curves represent empirical expressions which have been found to best fit the data.

The variation of G' with temperature was found to be adequately described by

$$G' = H \exp\left(-KT^{2/3}\right) \qquad \text{for } 0\,°\text{C} \le T \le 100\,°\text{C} \qquad (10)$$

where H and K are parameters for the material and depend upon strain and frequency. T is the temperature in °C and the power to which it is raised was determined as the value which enabled eqn. (10) to best describe the data. The accuracy of this expression is generally within ±5% over the ranges of temperature, strain amplitude and frequency considered here.

Figure 5 shows the variation of G' with log frequency for three strain amplitudes. The following equation produced the solid curve.

$$G' = B \log_{10} (f/f_0) \qquad \text{for } 0 \cdot 1 \text{ Hz} \leq f \leq 100 \text{ Hz} \qquad (11)$$

where B and f_0 are parameters for the material and depend upon temperature and strain amplitude. The departure of data from eqn. (11) is within $\pm 7\%$. The success of this function for describing the frequency dependence may prove valuable for deriving, by extrapolation, properties above 100 Hz where experimental difficulties are experienced in direct measurement.

The strong strain-amplitude dependence of modulus is apparent in Fig. 6 where it can be seen that the modulus rises rapidly with decreasing strain below 0·01. The extent of this dynamic strain-amplitude dependence will vary with the type and concentration of the carbon black. In Fig. 6, the solid lines represent two empirical curves that best fit the data. The expression

$$G' = G'_\infty + A \exp(-\lambda\gamma) \qquad (12)$$

was found to describe the strain dependence accurately for values of shear strain amplitude γ between 0·00125 and 0·01. G'_∞ is the limiting shear modulus at high strains and may be obtained to sufficient accuracy by measuring G' at $\gamma \geq 0\cdot 01$. A and λ are temperature- and frequency-dependent material parameters and may be determined from a plot of $\log(G' - G'_\infty)$ against γ. Despite the small range of strain amplitudes that this relationship covers, it is nevertheless useful due to the large variation of modulus over this range and its high accuracy (departure of data within $\pm 2\%$). At higher strain amplitudes, above 0·0025, the function

$$G' = P + Q\gamma^{-1/3} \qquad (13)$$

is found to be a better model than eqn. (12). Values for P and Q were obtained from a plot of G' against $\gamma^{-1/3}$. Like the parameters in the previous expression, P and Q are temperature- and frequency-dependent. This relationship was found to be accurate to within $\pm 5\%$ over the dynamic shear strain amplitude range $0\cdot 0025 \leq \gamma \leq 0\cdot 1$.

5.2. Prediction of G' values from a reduced set of measurements

In earlier work[5] a method was proposed for enabling the calculation of the value of G' at any frequency or strain from a limited series of experimental data. This method is based upon the assumption that the ratio of two values of G' at different strains γ_1 and γ_2 but the same frequency f_1 is independent of frequency and therefore equal to the ratio of moduli at any other frequency f_2 and the same strains. Thus

$$\frac{G'(f_1, \gamma_1)}{G'(f_1, \gamma_2)} = \frac{G'(f_2, \gamma_1)}{G'(f_2, \gamma_2)} \qquad (14)$$

If γ_2 and f_2 are chosen as reference values and γ_1 and f_1 allowed to take any values within the range of interest, then eqn. (14) can be expressed as

$$G'(f, \gamma) = \frac{G'(f, \gamma_r)G'(f_r, \gamma)}{G'(f_r, \gamma_r)} \qquad (15)$$

This means that the value for the shear modulus at any value of frequency or strain $G'(f, \gamma)$ may be calculated from a knowledge of the frequency dependence of G' at a reference strain level $G'(f, \gamma_r)$ and the strain dependence of G' at a reference frequency $G'(f_r, \gamma)$. The reference ranges chosen were: $G'(f_r, \gamma)$, $f_r = 3\cdot0$ Hz (as plotted in Fig. 6); and $G'(f, \gamma_r)$, $\gamma_r = 0\cdot01$ (as plotted in Fig. 5), which produced a value of $3\cdot28$ MN m^{-2} for $G'(f_r, \gamma_r)$. Table 2 shows the measured results over the full ranges of frequency and shear strain amplitude (in brackets), whilst the calculated values have been obtained from substitution of the reference values into eqn. (15). To illustrate the calculation method, G' at $1\cdot0$ Hz and 5% dynamic shear strain amplitude, $G'(1\cdot0, 0\cdot05)$, will be given from eqn. (15) by

$$G'(1\cdot0, 0\cdot05) = \frac{G'(1\cdot0, 0\cdot01) \times G'(3\cdot0, 0\cdot05)}{G'(3\cdot0, 0\cdot01)}$$

$$= \frac{3\cdot07 \times 1\cdot85}{3\cdot28}$$

$$= 1\cdot73 \text{ MN m}^{-2} \text{ (measured value } 1\cdot76 \text{ MN m}^{-2})$$

The predicted and experimental results agree within ±10% over the whole ranges, whilst the majority of the data are within ±5%.

Thus using G' data from reference values of temperature and dynamic shear strain amplitude it is possible to calculate the whole of Table 2 with sufficient accuracy for most applications.

TABLE 2

Comparison of G' Values $(\mathrm{MN\,m^{-2}})$ of a Carbon-Filled Rubber Obtained from Measurement (in Brackets) and from Calculation Using Only the Data for $G'(f_r, \gamma)$ and $G'(f, \gamma_r)$ in eqn. (15)

γ	f(Hz) $G'(f_r, \gamma)$							
	0·2	0·5	1·0	3·0	10	30	70	100
0·00125	4·10	4·86	5·47		6·75	7·67	8·80	9·10
	(4·64)	(5·22)	(5·63)	(5·84)	(6·88)	(7·92)	(8·95)	(9·36)
0·0025	3·73	4·43	4·98		6·15	6·99	8·01	8·29
	(4·12)	(4·68)	(5·09)	(5·32)	(6·25)	(7·18)	(8·15)	(8·49)
0·0050	3·13	3·71	4·17		5·15	5·86	6·72	6·95
	(3·31)	(3·83)	(4·23)	(4·46)	(5·21)	(5·96)	(6·81)	(7·07)
0·0075	2·66	3·15	3·55		4·38	4·98	5·71	5·90
	(2·72)	(3·20)	(3·57)	(3·79)	(4·41)	(5·03)	(5·76)	(5·96)
$G'(f, \gamma_r)$ 0·01	(2·30)	(2·73)	(3·07)	(3·28)	(3·79)	(4·31)	(4·94)	(5·11)
0·02	1·72	2·04	2·29		2·83	3·22	3·69	3·83
	(1·86)	(2·21)	(2·38)	(2·45)	(2·91)	(3·26)	(3·68)	(3·70)
0·05	1·30	1·54	1·73		2·14	2·43	2·79	2·88
	(1·41)	(1·64)	(1·76)	(1·85)	(2·14)	(2·33)	(2·62)	(2·69)
0·10	1·08	1·28	1·44		1·78	2·02	2·32	2·40
	(1·21)	(1·38)	(1·48)	(1·54)	(1·77)	(1·96)	(2·15)	(2·22)

A relationship analogous to eqn. (15) can also be postulated for describing the influence of temperature and strain on G', thus

$$G'(T, \gamma) = \frac{G'(T_r, \gamma)G'(T, \gamma_r)}{G'(T_r, \gamma_r)} \tag{16}$$

where T_r is a reference temperature. Table 3 shows measured values together with values (in brackets) calculated from the strain-amplitude-dependent data at a reference temperature of $G'(T_r, \gamma)$, $T_r = 25\,°\mathrm{C}$ (as plotted in Fig. 6), and the temperature-dependent data at a reference strain $G'(T, \gamma_r)$, $\gamma_r = 0·01$ (as plotted in Fig. 4). The calculated values are typically within 5% of the measured values, with slightly greater errors at the ends of the ranges.

The use of a similar method to model the combined frequency and temperature dependence of G' was less successful and it is concluded that the temperature and frequency dependence of G' are not

TABLE 3

Comparison of G' Values (MN m^{-2}) of a Carbon-Filled Rubber Obtained from Measurement (in Brackets) and from Calculation Using Only the Data for $G'(T_r, \gamma)$ and $G'(T, \gamma_r)$ in eqn. (16)

		Temperature (°C)			
		$G'(T_r, \gamma)$			
γ	*0*	*25*	*50*	*75*	*100*
0·00125	8·34			4·76	3·87	3·52
	(8·54)	(5·84)	(4·56)	(3·45)	(3·20)
0·0025	7·71			4·40	3·58	3·26
	(8·00)	(5·40)		(3·26)	(2·91)
0·005	6·50			3·71	3·02	2·74
	(6·68)	(4·55)		(2·82)	(2·55)
$G'(T, \gamma_r)$					
0·01	(5·00)	(3·50)	(2·85)	(2·32)	(2·11)
0·02	3·71			2·12	1·72	1·57
	(3·62)	(2·60)			(1·78)
0·05	2·79			1·59	1·29	1·18
	(2·70)	(1·95)	(1·78)	(1·51)	(1·47)
0·10	2·29			1·30	1·06	0·96
	(2·16)	(1·60)			(1·35)

separable in product form. This is expected from a knowledge of the origin of viscoelasticity in rubber materials[6,7] which requires a more complex time–temperature superposition model.

5.3. The shear loss behaviour

The shear loss factor of the filled rubber considered here shows only slight dependence on frequency (see Table 1). A description is therefore only necessary for the temperature and strain-amplitude dependence of loss factor, $d(T, \gamma)$. By analogy with eqn. (16), $d(T, \gamma)$ may be expressed in terms of the temperature dependence of d at a reference strain level $d(T, \gamma_r)$, and the strain dependence of d at a reference temperature $d(T_r, \gamma)$

$$d(T, \gamma) = \frac{d(T, \gamma_r)d(T_r, \gamma)}{d(T_r, \gamma_r)} \qquad (17)$$

Reference values of $T_r = 25\,°C$ and $\gamma_r = 0·01$ were chosen and calculated values of $d(T, \gamma)$ (in brackets) for a wide range of temperature

TABLE 4

Comparison of Loss Factors for a Carbon-Filled Rubber Obtained
from Measurement (in Brackets) and from Calculation Using Only
the Data for $d(T_r, \gamma)$ and $d(T, \gamma_r)$ in eqn. (17)

		Temperature (°C)			
		$d(T_r, \gamma)$			
γ	0	25	50	75	100
0·00125	0·23		0·14	0·10	0·09
	(0·20)	(0·17)	(0·12)	(0·09)	(0·10)
0·0025	0·25		0·15	0·10	0·10
	(0·24)	(0·185)		(0·11)	(0·10)
0·005	0·32		0·20	0·14	0·13
	(0·29)	(0·24)		(0·14)	(0·13)
$d(T, \gamma_r)$					
0·01	(0·40)	(0·30)	(0·25)	(0·17)	(0·16)
0·02	0·45		0·28	0·19	0·18
	(0·47)	(0·34)			(0·17)
0·05	0·41		0·26	0·18	0·17
	(0·45)	(0·31)	(0·24)	(0·19)	(0·16)
0·10	0·36		0·23	0·13	0·14
	(0·42)	(0·27)			(0·13)

and strain are compared with measured values in Table 4. The
agreement over the major part of the strain and temperature ranges
studied is within 10% and is comparable with the accuracy with
which results may be recorded.

6. DYNAMIC MECHANICAL PROPERTIES IN COMPRESSION AND COMBINED COMPRESSION AND SHEAR

For many applications such as engine mountings and other antivibration devices, rubber components are subjected to compressive or
combined compressive and shear deformations as depicted schematically in Fig. 9. For design requirements it is necessary to have a
knowledge of the dynamic stiffness of a component under such modes
of deformation and its variation with temperature, frequency and
strain amplitude. The chosen geometry of a sample will also affect its
dynamic stiffness. Thus additional testing needs to be carried out and

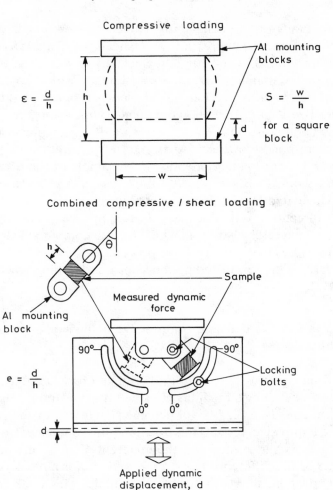

Fig. 9.   Schematic of load stages, illustrating shape factor S and compressive (ε) and combined compressive and shear (e) strains.

this produces a corresponding increase in the amount of data needed to characterise a materials behaviour. In earlier work,[5] procedures were developed for calculating the dynamic stiffness of a rubber block deformed under compression and combined compression and shear from suitable data obtained in simple shear. These procedures were used to obtain a set of calculated values which were compared with experimentally obtained values.

Measurements of the behaviour of filled rubbers under compression and combined compression and shear were made on square rubber blocks of side length w and thickness h with their end faces bonded onto aluminium blocks, in a similar fashion to the shear specimens described in Section 3.2. These were then loaded as shown in Fig. 9. The angle, θ, of the combined compression/shear loading stage was variable so that $\theta = 0°$ corresponds to compression, whilst $\theta = 90°$ corresponds to shear. At $0 < \theta < 90°$ a combination of compression and shear loading prevails. Each sample of the inverted 'V' arrangement was locked so that θ for each sample was identical. The strain is calculated by dividing the sample displacement in the vertical (loading) direction by its thickness. For compression loading, this is simply the compressive strain ε, whilst for the combined loading case this is the *component* strain e. Test procedures were as for the shear work described earlier, including the same conditioning schedule (see Section 3.2).

6.1. Analysis of the dynamic compressive behaviour of carbon-filled rubbers

Results from compression moduli determinations will differ from the material's true Young's modulus, E, since the constraint imposed by the bonded ends (see Fig. 9) will become significant for the geometry used in compressive tests. Thus the modulus obtained from these compressive tests is termed E_A, the apparent Young's modulus. The magnitude of the difference between E and E_A for a block of rubber will depend upon the geometry of the sample. This is described by the shape factor, S, defined in Fig. 9 for blocks of square cross-section and the larger the value for S, the greater the difference between E and E_A. The effect of sample geometry has been investigated for square blocks by making measurements over a range of strain amplitudes on samples having different shape factors, and the results are plotted in Fig. 10. The block is apparently stiffer at larger shape factors. The previous work on unfilled rubbers also derived a relationship between shear and compression moduli.[5] A necessary condition for this to be valid requires moduli to be compared at equivalent strain levels. It was shown that the equivalent shear strain γ to a compressive strain ε is given by

$$\gamma = \sqrt{(3)}\varepsilon \qquad (18)$$

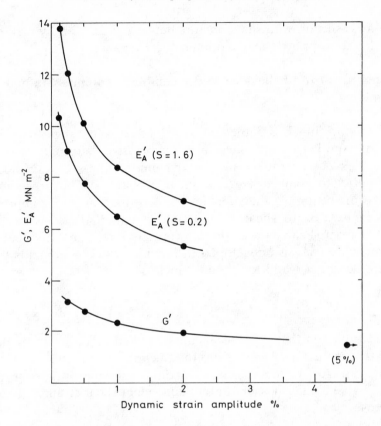

Fig. 10.   Variation of G' and E'_A with dynamic strain amplitude.

To clarify this point, a comparison between shear and compressive moduli can only be made if both are measured at *equivalent strains*, as given in eqn. (18). The expression previously derived for relating G' to E'_A for square blocks takes the form

$$E'_A(f, \varepsilon) = 3G'(f, \gamma)(1 + kS) \qquad (19)$$

where $\gamma = \sqrt{(3)}\varepsilon$, k is a material constant, and S is the shape factor. Plots of $E'_A(f, \varepsilon)$ against S were found to be linear for S in the range $1 \leq S \leq 2.5$ and the extrapolation of the straight line produced to $S = 0$ yielded an intercept of $3G'(f, \gamma)$, again, provided that E' and G' were obtained at equivalent strains. k is simply the slope of the line produced from plots of $E'_A(f, \varepsilon)$ against S.

Thus eqn. (19) can be used to calculate the apparent compressive modulus E_A at any frequency or strain amplitude for samples of square cross-section with shape factors between 1 and 2·5 by making use of the appropriate shear modulus data.

6.2. Analysis of the behaviour under combined compression and shear loading

When a rubber block is subjected to a load at some angle θ to its surface, it experiences both shear and compressive deformations (see Fig. 9). It is possible to relate the dynamic stiffness of the block $K^* = K' + iK''$ to compressive and shear properties. The procedure presented in Section 6.1 then enables the contribution from compression to be related to shear so that the dynamic stiffness may be calculated from shear data at the appropriate strain and frequency. This procedure was also developed in previous work[5] and yields the following equation (valid for square blocks)

$$K'(f, e, \theta) = \frac{w^2}{h} G'(f, \gamma)[\sin^2 \theta + 3 \cos^2 \theta (1 + kS)] \qquad (20)$$

It should be noted that e is the dynamic applied strain given by the ratio of the applied displacement to the thickness of the rubber block (see Fig. 9). As for compression loading, the appropriate shear modulus value is that measured at an equivalent shear strain γ, given in this case by

$$\gamma = e(\sin^2 \theta + 3 \cos^2 \theta)^{1/2} \qquad (21)$$

Figure 11 shows a comparison of the calculated modulus $M' = K'(h/w^2)$ (solid curve) of the sample under combined loading obtained from equivalent shear modulus data using eqn. (21) with measured apparent modulus values (data points) over a range of loading angle θ, and at different dynamic strain amplitudes. The close agreement between the experimental data and the theoretical predictions show that it is possible to determine the combined loading behaviour of V-blocks from just shear data using eqns. (20) and (21).

7. CONCLUSIONS

1. The variation of dynamic mechanical properties of carbon-filled rubbers with temperature, frequency and strain amp-

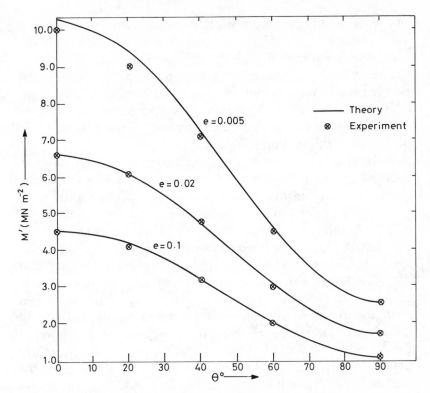

Fig. 11. Variation of apparent M' with loading angle θ for different dynamic strain amplitudes, e.

litude has been measured, and this variation was found to be quite significant. G', for one rubber, for example, ranged between 1·21 and 9·36 MN m^{-2} over the ranges of temperature, frequency and strain amplitude studied. The most notable feature for carbon-filled rubbers is the strong strain-amplitude dependence of dynamic moduli, particularly at low strains (less than 1%). Such large variations dictate that much data is needed in order to specify a materials behaviour.

2. A comparison was made between two methods of loss-factor determination to test whether the non-linearity invalidated phase angle measurements. However, for the materials studied no significant discrepancy was observed between the two methods.

3. Means of simplifying the presentation and reducing the quantity of measured data have been developed. An empirical curve-fitting exercise allows the dependence of G' upon temperature, frequency and strain amplitude to be described by simple relationships having at most three constants. This simplifies the presentation of a materials properties and also allows accurate interpolation between measured values which should be useful in design applications.

4. A method for determining the dynamic shear modulus G' over wide ranges of temperature, frequency and strain amplitude from chosen reference values was reviewed. This procedure works well for combinations of frequency and strain amplitude, and temperature and strain amplitude. However, the temperature and frequency dependences of G' were not separable in this way. The loss-factor data were also modelled similarly, and the variation of tan δ with temperature and strain amplitude could be satisfactorily described. Tan δ was found to have a negligible frequency dependence for the materials studied and therefore relationships involving frequency were not required.

5. Further streamlining of the presentation of material property data is also possible with the procedures reviewed here for obtaining the dynamic stiffness of square blocks in compression and combined compression and shear loading conditions. These procedures enable the calculation of such information from simple shear data and empirically determined shape factors, therefore obviating the need for a large set of measurements in compression and combined compression and shear.

ACKNOWLEDGEMENTS

Parts of the work described here were sponsored by British Leyland, Ford Motor Company, ICI Organics Division and the Transport and Road Research Laboratory. Our gratitude also extends to Dr D. M. Turner of the Avon Rubber Company for useful discussions and the provision of materials for testing.

REFERENCES

1. Read, B. E. and Dean, G. D. (1978). *The Determination of Dynamic Properties of Polymers and Composites*, Bristol, Adam Hilger Ltd.
2. Payne, A. R. (1962). *J. Appl. Polym. Sci.*, **6**(19), 57–63.
3. Freakley, P. K. and Payne, A. R. (1978). *Theory and Practice of Engineering with Rubber*, London, Applied Science Publishers Ltd.
4. BS 903: Part A24 (1976). *Methods of Testing Vulcanized Rubber.*
5. Johnson, A. F. and Dean, G. D. In preparation.
6. Ferry, J. D. (1961). *Viscoelastic Properties of Polymers*, New York, John Wiley and Sons Inc.
7. Cowie, J. M. G. (1973). *Polymers: Chemistry and Physics of Modern Materials*, Aylesbury, International Textbook Co. Ltd.

Polymer Testing **4** (1984) 251

The Time and Temperature Dependence of Fracture Toughness

D. R. Moore

Imperial Chemical Industries PLC, Petrochemicals and Plastics Division, Wilton Centre, PO Box 80, Wilton, Middlesbrough, Cleveland TS6 8JE, UK

SUMMARY

The toughness of thermoplastics can be measured in a number of ways. An intrinsic measure emerges in terms of fracture toughness. More precisely, a critical value for stress field intensity factor (K_c) in combination with a yield stress can be shown to be a useful measure of ductility. Such an approach can be helpful for material comparison, development and selection, and also in design considerations.

It is apparent that K_c is a viscoelastic function for thermoplastics. Consequently, it is important to measure its time and temperature dependence and appropriate techniques have been developed. This is particularly the case for times under load in the range 1 ms to 10 s.

There is also a need to conduct a careful interpretation of measured values for K_c. This is necessary because K_c can be dependent on both specimen geometry and on the extent of the ductility in the region of a growing crack. Various criteria are available for resolving such dependence.

These considerations were exemplified by work on a range of thermoplastics and for a number of test conditions.

Polymer Testing **4** (1984) 253–272

Measurement and Analysis of Slow Crack Growth in a Viscoelastic Material

G. D. Dean and L. N. McCartney

National Physical Laboratory Teddington, Middlesex TW11 0LW, UK

SUMMARY

Apparatus for studying slow crack growth in a polymer is described. A theoretical analysis is outlined which relates the speed of growth to the stress intensity factor and material properties, and demonstrates how the viscoelastic behaviour of the polymer is responsible for slow crack growth. The analysis predicts a threshold level of stress intensity below which crack growth under static load should not occur. Experimental data on slow crack growth in two grades of PVC have been obtained for comparison with theoretical predictions. Difficulties experienced with the collection of stable growth data on these materials are associated with the occurrence of transient growth and the reluctance of the crack to advance at the surfaces of the test piece. The theory is observed to describe slow growth in a small range of stress intensity close to threshold only. At higher levels, experimental growth rates for both PVC grades are much lower than theoretical values and the range of stress intensity giving stable growth is greater than that predicted. A modification to the theoretical analysis is considered which attempts to model more accurately the behaviour of the plastic zone material, and preliminary calculations have indicated that this produces a better description of experimental data.

1. INTRODUCTION

The observation of brittle failures in most engineering plastics after long times under load has stimulated research on the subjects of

253

crack initiation and growth in these materials. These subjects are difficult to study because the mechanisms involved are operative over long periods of time. In addition, the specification of materials behaviour is complicated by the fact that crack initiation times and growth rates are influenced in complex fashions by a wide range of factors. Polymer structure, temperature, the presence of alternating loads and particular chemical environments are just some of the factors that have been observed to affect the nature and kinetics of brittle fracture processes.

Materials characterisation thus involves a great deal of time-consuming testing and some consideration should be given to the selection of test methods and analyses to ensure relevant and meaningful results. The availability of procedures for interpreting data would reduce the quantity of testing needed to describe fracture behaviour and enable the characteristics of a polymer to be presented in a form suitable for materials comparison or design. A substantial amount of work has been reported in the literature on this subject,[1-3] but there is scope for further effort aimed at handling the testing and analysis of a wider range of polymers under a greater variety of loading situations.

In this paper, a fracture mechanics analysis is outlined for describing crack growth in a viscoelastic material. It demonstrates that slow stable growth is a consequence of time dependence in the modulus of the material and relates crack growth rates to the applied stress intensity. The theory therefore appears attractive for the analysis and interpretation of slow crack growth experiments. It also relates the minimum or threshold level of stress intensity, below which crack growth should not occur, to quantities that can be measured in short or simple test procedures. This threshold stress intensity factor would constitute a useful and relevant parameter for defining the resistance of a polymer to slow crack growth.

In order to assess the validity of this theory for describing slow crack growth in a polymer, theoretical predictions are compared with experimental crack growth data obtained on two grades of PVC. It is apparent that there are certain difficulties that can arise in the collection of stable growth data on these materials. In addition there are several features of the growth behaviour which are not predicted by the theory. An interpretation of these features is presented, and, in the light of this, improvements to the theoretical model are considered. In its present form, the proposed analysis is complicated but should be considered as a basis from which valid and more workable theories could be developed.

2. DERIVATION OF THE CRACK GROWTH LAW FOR A VISCOELASTIC MATERIAL

An outline of the steps leading to a derivation of the growth law is given in this section; the mathematical details of the analysis are presented elsewhere,[4,5] Figure 1 shows a crack of length $c(t)$ in a polymer that is loaded in the direction x_2 normal to the plane of the crack. A line plastic zone ahead of the crack tip is depicted as a single craze having a length $R(t) = a(t) - c(t)$ and is assumed at present to support a stress σ_p which is independent of time and position in the zone. Under slow crack growth the crack advances through the zone and new plastically deformed material is incorporated simultaneously into the zone at its tip. The rate at which work is done on the plastic zone by the stress field in the remainder of the polymer is then balanced by the rate at which energy is dissipated in the zone through plastic deformation and crack growth. Equating these quantities yields the following local energy balance equation

$$\int_{c(t)}^{a(t)} \sigma_p \frac{\partial \Delta u_2(x_1, t)}{\partial t} \, dx_1 = 2\Gamma \dot{c}(t) \qquad (1)$$

where $\dot{c}(t)$ is the speed of growth, $\Delta u_2(x_1, t)$ is the discontinuity of the component u_2 representing the displacement of the opposite surfaces of the zone, and Γ is the fracture energy per unit area. The quantity Δu_2 may be expressed in terms of x_1, t, R and the tensile creep function $J(t)$ of the polymer. The magnitude of the stress field around

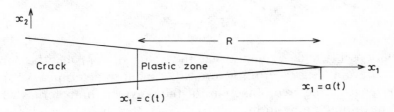

Fig. 1. Schematic diagram of the region around the tip of a crack in a material.

the crack increases as the plastic zone is approached and it must be bounded at the point $x_1 = a(t)$, $x_2 = 0$. This requirement leads to the following relationship for the zone length R (valid only for the line plastic zone model under discussion)

$$R(t) = \frac{\pi}{8} \frac{K(t)^2}{\sigma_p^2} \tag{2}$$

where $K(t)$ is the stress intensity factor and defines the magnitude of the stress field around the crack tip. $K(t)$ can be determined from a knowledge of the applied load P and the crack length c through relationships of the form

$$K(t) = Pc(t)^{1/2} Y \tag{3}$$

The parameter Y is governed by the size and geometry of the loaded material and the crack length. Expressions exist[6] enabling Y to be calculated for a wide range of common geometries. If the relationships for Δu_2 and R are substituted into eqn. (1), an expression for the crack speed $\dot{c}$ may be derived in terms of K and the material properties $J(t)$, Γ and σ_p. This is the crack growth law for the polymer.

The theory demonstrates that slow crack growth is a direct consequence of the viscoelastic behaviour of a material and stipulates that growth will only take place if the stress intensity lies between a minimum value K_{TH} and a maximum value K_c. If the compliance function $J(t)$ approaches a limiting value $J(\infty)$ at long times, then the threshold stress intensity is given by

$$K_{TH}^2 = \frac{2\Gamma}{J(\infty)} \tag{4}$$

The upper bound K_c, for which $\dot{c} \to \infty$ as $K \to K_c$, is governed by the zero-time or elastic limit $J(0)$ to $J(t)$ and is given by

$$K_c^2 = \frac{2\Gamma}{J(0)} \tag{5}$$

Equations (4) and (5) imply that the value of K_{TH} may be simply derived from a short-term fracture experiment to determine K_c and a knowledge of the creep function for the polymer. In materials which do not exhibit a long-term limit to $J(t)$, no threshold to crack growth is predicted. The implication then is that, if a long-time value for $J(t)$

were substituted into eqn. (4), the calculated value of K_{TH} would correspond to crack growth at a very slow speed which, for most practical purposes, constitutes a threshold level.

In order to assess the validity of this theory for describing slow crack growth in a polymer, predicted growth laws have been compared with experimental data obtained for two grades of PVC.

3. MATERIALS

Two transparent grades of PVC were obtained from ICI PLC. One of these was extruded sheet of about 6·4 mm thickness manufactured under the trade name Darvic®. The other was compression moulded material, also about 6·4 mm thick, having a chemical specification closer to that used in engineering grades of PVC with the change of ingredient necessary to render the material transparent. These materials are designated PVC-D and PVC-H, respectively. Material H showed no birefringence. In the sheet of D material, the in-plane birefringence varied with position in the sheet from 0 to about 2×10^{-4}. This birefringence could only be removed by heating above 180 °C. It was concluded, therefore, that its origin was due to molecular orientation maintained at temperatures above T_g by microcrystalline regions acting as cross-links in the molecular chain network. No thermal treatments were applied to the samples used for the crack growth studies. All materials are therefore as-received, and the majority of data has been obtained at least 2 years after procurement.

4. EXPERIMENTAL

In this section, apparatus is described for obtaining data on slow crack growth. Theoretical growth rate predictions require information on the creep compliance function $J(t)$ for the polymer, the fracture energy Γ and the plastic zone stress σ_p. It is proposed to determine Γ from measurement of K_c, using eqn. (5), and σ_p from measurement of R using eqn. (2). Test procedures are also described for evaluating these quantities experimentally.

4.1. Crack growth studies

The objective of the crack growth studies is to determine how the velocity of slow stable crack growth is related to the stress intensity. Experiments are carried out on specimens in which a crack has been artificially introduced and for which the stress intensity at the tip of the crack arising from an applied load can be calculated.

4.1.1. *Specimen geometry*

All measurements of crack growth were made on compact tension specimens having the geometry depicted in Fig. 2. For a specimen of width W and thickness B having a crack of length c and under a load P, the stress intensity K is calculated thus[6]

$$K = \frac{PY\sqrt{c}}{BW} \tag{6}$$

where Y is a function of c/W. For $0\cdot3 < (c/W) < 0\cdot7$, it is stated that

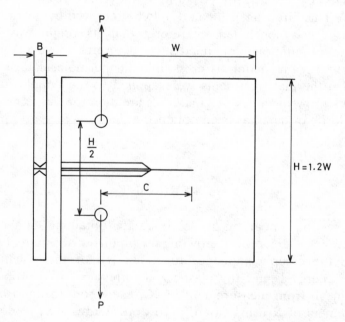

Fig. 2. Compact tension specimen geometry for crack growth studies.

Y is given to an accuracy of 1% by

$$Y = 29 \cdot 6 - 186\left(\frac{c}{W}\right) + 656\left(\frac{c}{W}\right)^2 - 1017\left(\frac{c}{W}\right)^3 + 639\left(\frac{c}{W}\right)^4 \quad (7)$$

The notch is introduced by fly-cutting both surfaces so that the cuts meet in the centre of the specimen. The preparation of a sharp crack may be readily achieved in PVC by fatigue loading. The crack is initiated by inserting a small razor cut at the tip of the notch and fatigue loading until the crack has growth out of the chevron produced by the fly-cutting.

4.1.2. Apparatus

Crack growth investigations under constant load have been carried out on simple apparatus consisting of a cross-head from which the compact tension sample is suspended via one of the pin holes in the sample. The load P is generated by the application of weights to a pan hanging from the other hole. The crack length is determined using a low-power travelling microscope viewing the crack tip at an angle of about 45° to the plane of the crack. Changes in crack tip position with time can usually be recorded to ±0·01 mm using a micrometer eyepiece or the vernier scale on the microscope. To eliminate random movements of the test piece with respect to the foundation of the microscope, it is necessary to record the position of a reference mark on the specimen (usually a fine scratch) alongside each crack tip reading.

Since the load remains constant, the stress intensity will increase as the crack advances. Reasonable accuracy in crack speed measurements can, however, usually be achieved for small increases of crack length ($\simeq 0 \cdot 5$ mm). These will give rise to only small changes in Y, so K is essentially constant during the crack speed determination. The load, and hence the stress intensity, is then raised incrementally in order to obtain further crack speed data over the range of K responsible for stable crack growth.

One disadvantage of constant load testing arises when there is a possibility that the crack growth rate might increase substantially as this could lead to specimen failure before measurements of crack position can be made and the load reduced. This situation arises with the testing of materials exhibiting transient growth as discussed in Section 5. It is then more suitable to test under constant applied displacement so that as the crack advances the stress intensity decreases. For this purpose a simple loading assembly has been constructed whereby the specimen is connected in series with a force transducer to a threaded shaft which enables a static displacement to be applied. The other end of the sample can be connected to a rigid foundation. A simple fatigue facility becomes available if this foundation is replaced by a source of dynamic displacement achieved, for example, by means of an eccentric shaft driven by an electric motor. The apparatus then enables sharp cracks to be prepared by dynamic loading.

The amount by which the stress intensity falls with crack growth in a constant displacement test can be reduced by inserting an elastic (metal) ring in series with the specimen. As the compliance of the specimen increases due to crack advance, the displacement across the specimen will increase by an amount dependent upon the magnitude of the compliance of the ring in comparison with that of the specimen. It is possible to select a compliance for the ring such that the stress intensity is held almost constant for a substantial increase in crack length.

4.2. Measurement of plastic zone length

The crack surface appears very smooth in a specimen prepared by low-amplitude fatigue cycling ($K_{max} \leq 0.2\ \mathrm{MN\ m^{-3/2}}$). If a load is then applied giving rise to a stress intensity above the fatigue maximum but below that which causes transient crack growth, then the size of the plastic zone resulting from this load may be measured since it appears rougher. The zone length R is observed to grow with time under load. Data on the increase of zone length with time for both grades of PVC are recorded in Figs 3 and 4 at different values of the stress intensity. Prior to each set of measurements, the sample was fatigued to prepare a new crack tip having a relatively small plastic zone again.

4.3. Determination of creep compliance

Creep studies were carried out on rectangular strips of uniform cross-sectional area A held under constant tensile load P. The longitudinal strain $\varepsilon(t)$ was measured with time under load, and the

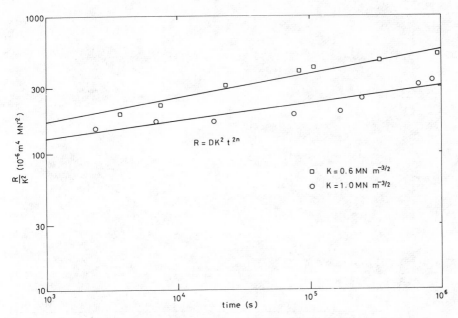

Fig. 3. Variation of plastic zone length R with time t under load plotted on logarithmic axes of R/K^2 against t. Darvic material.

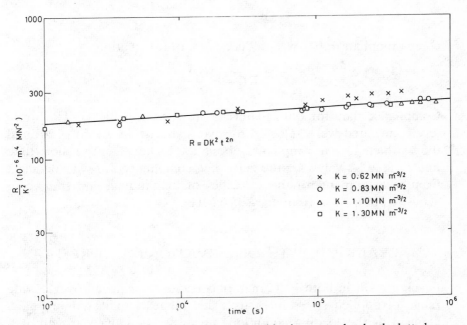

Fig. 4. Variation of plastic zone length R with time t under load plotted on logarithmic axes of R/K^2 against t. PVC material H.

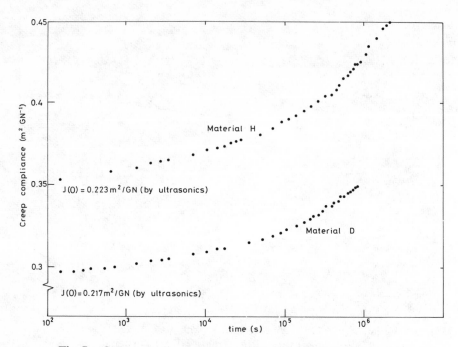

Fig. 5. Creep compliance against log (time) for both PVC materials.

creep compliance $J(t)$ was deduced from the equation

$$J(t) = \frac{\varepsilon(t)A}{P} \tag{8}$$

Compliance data for both materials are shown in Fig. 5. The stress levels indicated were selected to give a short-time strain of around 0·5% which gave a compliance about 1% higher than the short-time linear value. A value for the zero-time compliance $J(0)$ was deduced from measurements of the velocities of longitudinal and transverse acoustic pulses at a frequency of 5 MHz.

5. CRACK GROWTH OBSERVATIONS AND RESULTS

Crack growth in both PVC materials could be achieved over a wide range of applied stress intensity. This range is indicated in Fig. 6 along with the loading conditions necessary to induce growth at a

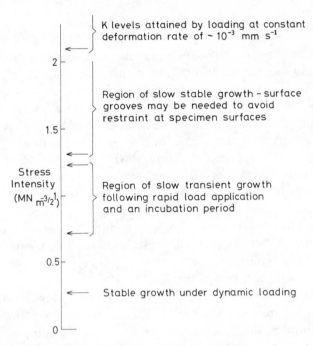

K levels attained by loading at constant deformation rate of ~ 10^{-3} mm s^{-1}

2 —

Region of slow stable growth – surface grooves may be needed to avoid restraint at specimen surfaces

1.5 —

Stress Intensity (MN m$^{-3/2}$)

1 —

Region of slow transient growth following rapid load application and an incubation period

0.5 —

Stable growth under dynamic loading

0 —

Fig. 6. Range of stress intensity K for which slow crack growth can be induced in both grades of PVC.

particular K level. Under cyclic loading at a frequency of about 1 Hz, slow stable growth can be achieved at dynamic K amplitudes of around 0.25 MN m$^{-3/2}$. This loading schedule was used to generate a sharp crack in a specimen prior to static load experiments. If a stress intensity in the region of 1 MN m$^{-3/2}$ was applied fairly rapidly (within 10 s) to a sample in which a fatigue crack had been induced, crack advance in both PVC grades would usually follow an incubation period (about 1 h). This growth was a transient phenomenon and, as long as the applied K was not too high, the growth rate would decrease to zero.

Transient growth could be substantially reduced if the stress intensity level was reached by loading in small increments with around 24 h separating each load application. Under these conditions, comparable K values could be achieved with only marginal transient crack growth taking place and only then in the central region of the sample. Consequently, the crack tip profile would become increasingly curved after each load application. If this incremental loading procedure was continued, slow stable crack growth could be induced in material H at stress intensities of about $1 \cdot 3 \, \text{MN} \, \text{m}^{-3/2}$ and above. In the Darvic material, slow stable crack advance was unusual even at K values approaching $2 \, \text{MN} \, \text{m}^{-3/2}$. In this material, the plastic zone, which at these K levels appeared to consist of a bunch of crazes, was apparently larger, both in length and thickness, at the surfaces of a sample than near the centre. In an attempt to relieve what appeared to be a constraining influence of the sample surface upon crack advance in this material, V-grooves were inserted to a depth of $0 \cdot 5 \, \text{mm}$ (sample thickness $= 6 \cdot 4 \, \text{mm}$) in each surface to coincide with the plane of the crack. It was then possible to achieve slow stable growth at values of stress intensity around $1 \cdot 4 \, \text{MN} \, \text{m}^{-3/2}$ where no continuous growth was generally observed in the ungrooved samples. In those ungrooved samples which showed slow continuous growth, the influence of inserting the grooves was apparently to raise the speed of growth at a particular K.

Data relating the crack speed for stable growth in material D to the stress intensity are shown in Fig. 7. These data were obtained on four different samples, three possessing grooves and one ungrooved. Each set of data exhibits a sharp onset of crack growth. With three of the samples this is followed by a range of stress intensity for which the growth rate was approximately constant after which the speed increased rapidly again prior to failure of the sample. In the other sample this period of sustained slow growth was absent. Data at growth rates higher than those recorded here ($>10^{-5} \, \text{mm} \, \text{s}^{-1}$) could not be obtained. It is believed that once these speeds were reached the crack would accelerate even under constant or slowly reducing K and sample failure would soon result. The apparent scatter in data arises since the K range for stable growth is slightly different for each sample. The dotted line in Fig. 7 demonstrates the trend in behaviour and refers to one of the samples. No attempt has been made to allow for the presence of the grooves in the calculation of K values.

Data showing the variation of crack speed with stress intensity for material H are shown in Fig. 8. These values were obtained on two samples represented by different symbols. Data at the lower values of K were obtained on one of the samples by progressively decreasing the stress intensity until the crack stopped growing. The scatter in velocity is real and reflects a variation that could be obtained daily. The data generally represent the average value for several days growth.

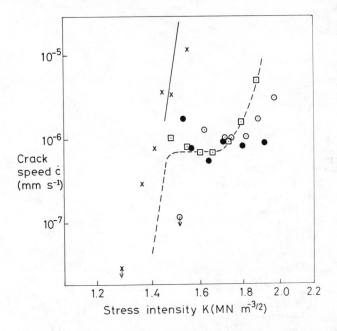

Fig. 7. Variation of crack velocity $\dot{c}$ with stress intensity K for stable growth in Darvic material. $\odot$, date on a sample without side grooves; other symbols refer to data on grooved samples. – – – represents the typical trend in behaviour; and ——— the theoretical prediction.

Tests at constant deformation rate of the order of 10^{-3} mm s^{-1} lead to failure of a specimen in a time-scale of around 10^2 s. Contributions from the phenomena of incubation and slow growth to the fracture event are then minimised or avoided. Samples of the Darvic material would fail by the rapid advance of the crack at a stress intensity of about $2 \cdot 2$ MN m$^{-3/2}$. The H material was able to support a stress intensity of greater than 6 MN m$^{-3/2}$ without rupture or obvious crack growth by which time large-scale plastic deformation had developed at the crack tip and the test was abandoned.

6. ANALYSIS OF EXPERIMENTAL RESULTS

The creep compliance data in Fig. 5 may be accurately described by the function

$$J(t) = J(0) + At^\lambda \tag{9}$$

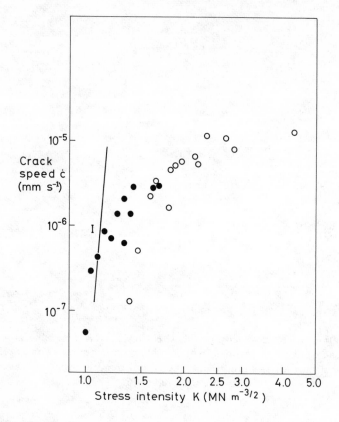

Fig. 8. Variation of crack velocity $\dot{c}$ with stress intensity K for stable growth in PVC material H.

where $J(0)$ is the instantaneous or elastic component of the compliance determined by ultrasonic wave velocity measurements, and A and λ are constants for the material. Values for $J(0)$, A and λ for each PVC material are given in Table 1.

The observation that the length of the plastic zone increases with time under constant K may be explained by proposing a time-dependent value for the plastic zone stress. The data in Figs 3 and 4 indicate that $\log R/K^2$ is a linear function of $\log t$ implying that

$$\sigma_p = \sigma_o t^{-n} \tag{10}$$

at large values of t, where σ_o and n are constants for the material.

TABLE 1
Experimental Values for the Creep Parameters[a]
in Eqn. 9 for both PVC Materials

Parameter	PVC material type	
	D	H
$J(0)$	0·217	0·224
A	0·131	0·132
λ	0·031	0·062

[a] Units: GN, m, h.

Equation (2) then becomes

$$R = DK^2 t^{2n} \tag{11}$$

where $D = \pi/8\sigma_o^2$ for the line plastic zone model. Values for D and n for each material obtained empirically are given in Table 2.

TABLE 2
Experimental Values for the Parameters[a] Defin-
ing Plastic Zone Growth in Eqn. 11 for both PVC
Materials

Parameter	PVC material type	
	D	H
D	50×10^{-6}	120×10^{-6}
$2n$	0·15	0·05

[a] Units: MN, m, s.

The results for zone growth were obtained for cracks which were stationary. For a steadily growing crack, it is assumed for the moment that eqn. (11) remains valid, the parameter t now being interpreted as the time taken for the crack to grow through the length of the zone. Thus $t = R/\dot{c}$ and eqn. (2) is finally replaced by

$$R = (DK^2)^{1/1-2n} \dot{c}^{2n/(2n-1)} \tag{12}$$

The unstable failure observed in the Darvic material in a constant deformation rate test throws some doubt on whether this experiment yields a valid result for K_c. Furthermore, since no meaningful value for K_c could be determined experimentally for material H, it has been necessary to select Γ (or K_c) of both grades by achieving the best fit of theoretical predictions of the crack growth law to experimental data. This analysis is a complicated exercise which has been described in detail elsewhere.[5] The resulting growth curves are represented by the solid lines in Figs 7 and 8. The gradient of these lines is not influenced by the magnitude of the selected Γ value; the latter only shifts the line parallel to the K axis. The positions of the best-fit lines were thus chosen to coincide with data where the gradient was similar. The termination of the theoretical lines at low crack speeds is caused by the limited time-scale over which creep data was collected.

7.   DISCUSSION

It is apparent from a comparison of the theoretical crack growth predictions with experimental data that, at best, the theory is only able to describe crack growth near threshold. At higher values of K, the crack speed rises less sharply than predicted leading to slow growth over a wider range of K than implied by eqns. (4) and (5). The reasons behind this poor agreement could stem from inadequacies in the model when applied to PVC or limitations on the validity of the experimental data presented here. Some of the possible causes are considered later in this section but an attempt is made first to give an interpretation of crack growth behaviour in the materials studied.

7.1. An interpretation of the salient features

It is clear that the resistance to crack growth of the grades of PVC studied increases with time under load and, in comparison with the theory, with increasing load. This behaviour would appear to be associated with the development of the structure of the crazed material in the zone ahead of the crack. This material is known to be highly non-linear and viscoelastic,[7] and the fact that structural changes take place over a time-scale of days is demonstrated by the results of measurements on plastic zone length shown in Figs 3 and 4. The larger size of the zone at the sample surfaces, coupled with the reluctance sometimes for the crack to advance at the surface, also bears evidence to the belief that structural changes in the zone can lead to a situation in the sample which is more resistant to crack advance. The properties of the zone could contribute directly to the enhanced toughness or indirectly through a change in the stress field immediately ahead of the crack.

The onset of transient growth, whereby the crack advances when the load is applied relatively rapidly, would take place before the zone had reached an equilibrium structure. If the crack tip speed under transient growth were such that the age of the zone ($\sim R/\dot{c}$) was greater than the time required for the zone to develop a more resistant structure, then the crack would decelerate. The converse situation, in which a weak zone structure is established or maintained by virtue of high crack speeds, could go some way towards explaining the unstable growth observed in the Darvic material at the higher levels of K in slow growth experiments and in constant deformation rate tests.

7.2. Stable growth and the theoretical model

The theoretical growth laws are appropriate only for a crack having a single craze at the crack tip as depicted schematically in Fig. 1. Observation suggests that, whilst a single craze may exist at small values of the stress intensity ($K < 0.3\,\text{MN m}^{-3/2}$), multiple crazing is observed at the higher values used in these experiments. The relationship used for the displacement discontinuity Δu_2 which is substituted into the energy balance eqn. (1) will be different in this multiple crazing situation. Further improvements to the model should also be made regarding its ability to describe plastic deformation ahead of the crack. It is known that craze material exhibits the behaviour of a non-linear, viscoelastic solid implying that the quantity σ_p in eqn. (1) should be replaced by a constitutive relationship demonstrating the dependence of σ_p upon deformation history. Although the experiment has indicated that σ_p can be described by eqn. (10), this was shown in Figs 3 and 4 to be true for a limited range of K well below those levels for which experimental and theoretical crack growth data depart. The time dependence in plastic zone behaviour may also lead to a relaxation of the stress field around the crack tip (blunting?) so that its magnitude *local* to the tip may no longer be characterised by the calculated stress intensity value given by eqn. (6).

A modification to the model has been considered, and preliminary calculations have shown that, by a suitable choice of constitutive law for the craze zone, a crack growth curve resembling that obtained experimentally for material H can be predicted. The larger gradient at lower stress intensities is predicted including a threshold value for K. If the effective fracture energy is calculated using this model, it is found to be related to the magnitude of K and the time. The model is therefore consistent with the interpretation of material behaviour outlined in Section 7.1. It also demonstrates that time dependence in the mechanical properties of the plastic zone is an additional factor responsible for the phenomenon of slow crack growth in a material.

7.3. The presence of plane stress

The observation that the surfaces of a sample appear to have a restraining influence on crack advance, especially in material D, has been noted in Section 5. This restraint is almost certainly associated with the presence of a region of plane stress in the zone near a surface in which presumably the properties and structure of the crazed material are different from those in the zone nearer the sample centre. It has been reported in the literature[8,9] that the presence of even small plane stress regions can influence the maximum stress intensity that a sample can sustain. A reduction in the restraint imposed by the sample surfaces can be achieved by grooving the surfaces. One effect of this is to allow a hitherto stationary crack to advance at a given K level. It might be concluded that the threshold for growth under plane stress is higher than that under plane strain and, until the former is reached, no continuous growth will take place.

It is not clear how the plane stress regions will influence the speed of slow stable crack growth. It would seem reasonable that the extent of any influence would depend upon the crack growth characteristics under plane stress in comparison with those under plane strain and the width of the plane stress zones compared with the sample thickness. The width of the plane stress zones is similar in magnitude to the length of the plastic zone under plane stress. No direct measurements have been made of this, but the plane strain zone length can be calculated using the information presented in eqn. (11) and Table 2. It is apparent that at a stress intensity of around

2 MN m$^{-3/2}$ and after a time of 10^6 s the plane stress regions in both materials are likely to be a significant fraction of the sample thickness ($\approx 6\cdot4$ mm). A transition from a predominantly plane strain condition ahead of the crack to plane stress might therefore be responsible for the change of gradient in the growth data for material H (Fig. 8). This is unlikely to be an explanation for the similar phenomenon observed in Darvic material since in this material this change is very sharp. Furthermore, the side grooves in the Darvic samples should have reduced the influence of the plane stress field.

8. CONCLUSIONS

A rigorous theoretical treatment is described which demonstrates that, for a viscoelastic material (by virtue of its time-dependent modulus), there is a range of stress intensity for which slow stable crack growth is possible. Using the theory, a growth law is derived which relates the speed of growth to the magnitude of the stress intensity.

Experimental studies have shown that slow stable growth in PVC is difficult to establish and can be obscured by transient growth phenomena. Stable growth occurs over a wider range of stress intensity than that predicted by the theoretical model. This appears to be associated with an apparent increase in the toughness of the materials studied at higher K values which is manifested by a marked decrease in the gradient of the crack growth curve. Although the theory has had limited success in describing slow growth behaviour in PVC, it may be satisfactory for other polymers.

The following additional observations also cannot be explained by the steady state theory described. The onset of growth in PVC depends upon how quickly the load is applied and the growth is strongly influenced by cyclic loading. In one of the grades of PVC, once the speed of growth reached a critical value the crack tip would accelerate even at a constant stress intensity. Stable growth can be achieved by the slow application of load which may take days if transient effects are to be avoided. The onset and speed of growth can be restrained by the surfaces of the sample.

A preliminary extension of the theory has shown that if the material in the plastic zone ahead of the crack behaves like a non-linear viscoelastic solid then this can have a substantial influence on growth behaviour. In such a material, the time dependence of the properties of the bulk polymer might be of less significance. This modification to the model is consistent with a situation in which the toughness of the polymer depends on the structure of the material in the plastic zone ahead of the crack. In order to explain the experimental observations, this structure would have to be dependent on the time under load and the magnitude and multi-axiality of the stress field to which the zone is subjected; requirements which are quite reasonable if the zone consists predominantly of crazed material.

Some problem areas have been identified and require further investigation before a test procedure can be defined for evaluating the slow crack growth behaviour of certain polymers. Most notable of these are transient growth phenomena and the restraining influence of the sample surfaces. In the context of the latter, the relevance of results on samples with side grooves or with a highly curved crack front needs to be established.

REFERENCES

1. Williams, J. G. (1978). *Advances in Polymer Science*, **27**, 67–120.
2. Andrews, E. H. (Ed.) (1979). *Developments in Polymer Fracture—1*, London, Applied Science Publishers Ltd.
3. Bucknall, C. B. (1977). *Toughened Plastics*, London, Applied Science Publishers Ltd.
4. McCartney, L. N. (1979). *Int. J. Fracture*, **15**, 31–40.
5. Dean, G. D., McCartney, L. N., Cooper, P. M. and Golding, S. L. (1982). National Physical Laboratory Internal Report No. DMA (A) 57, Teddington.
6. Rooke, D. P. and Cartwright, D. J. (1976). *Compendium of Stress Intensity Factors*, London, HMSO.
7. Kambour, R. P. (1973). *J. Poly. Sci.*, **7**, 1–154.
8. Fernando, P. L. and Williams, J. G. (1980). *Poly. Engng. Sci.*, **20**, 215–20.
9. Pitman, G. L. and Ward, I. M. (1979). *Polymer*, **20**, 895–902.

Polymer Testing **4** (1984) 273–288

Multiaxial Mechanical Testing of Plastics

C. C. Lawrence

North East London Polytechnic, Manufacturing Studies, Longbridge Road, Dagenham, Essex, UK

SUMMARY

This paper outlines the many problems associated with the testing of plastics and suggests a method of minimising these problems by the use of multiaxial testing techniques. It extends the multiaxial approach into the determination of the fatigue loading characteristics of a plastic. General design criteria of such machines, as well as details of a family of machines that have been developed for that purpose, are given.

1. INTRODUCTION

Industrial design has now reached the stage where plastic materials are considered along with other materials on an 'equal' basis. However, there is still the tendency for designers to apply traditional techniques that require data from conventional test methods that were developed for the testing of metals. Not only does this approach disregard many of the most favorable aspects of plastic materials, but some information such as standard fatigue (S–N) data is effectively unobtainable. The materials dynamic behaviour (fatigue) is none the less a very important aspect of the design of engineering components and some estimation of the fatigue performance of the material must

Polymer Testing 0142-9418/84/$03·00 © Elsevier Applied Science Publishers Ltd, England, 1984. Printed in Northern Ireland

be found. This paper sets out to show:

1. That the inherent characteristics of polymeric materials make the collection of conventional data impractical.
2. That an alternative technique is load simulation which will lead to a greatly improved understanding of the materials performance under complex loading.
3. The design criteria of test equipment that is to be used for multiaxial loading. Some of the unusual features of a family of multiaxial machines designed to meet this requirement are discussed.

2. CONVENTIONAL DATA

Conventional fatigue data are usually presented in the form of stress versus (log) cycles to failure (S–N) curves. These curves are derived from a great many tests on a closely specified material. Fatigue data are subject to considerable scatter hence the requirement for many specimens and for long life estimation. These tests are normally extended to include lives in excess of 10^7 cycles. Principally because of time constraints these results are obtained by high-frequency tests, typically 50 Hz (approximate 2 days for 10^7 cycles). It is essential that the loading of the specimen is kept below a level that will avoid excessive heating. However, with plastics the viscoelastic behaviour and low thermal conductivity combine to give considerable heating at quite low loading if the frequency is of the order of the 50 Hz cited above. Whilst the shortcoming is the primary reason for the lack of fatigue data in plastics, there are other factors that have precluded the collection of standard fatigue data. Although these limiting factors are closely linked, they will be separated into those parameters that are related to the physical characteristics of the material and the working environment. The operational environment can only be considered in detail with respect to the particular set of operating conditions, such as temperature and loading pattern, but, with the limited operational temperature of most commercially available plastics, temperature and severity of loading are of extreme importance.

3. PROPERTIES INHERENT IN THE MATERIAL

Metals are manufactured to tight specifications and consequently any generalised data can be applied without exception. Plastic materials that have been manufactured within the 'specification' none the less can have quite dramatic variations in properties due to the actual manufacturing processes. Some of the variations in properties are due to factors within the plastic itself, others are induced externally. Examples of these modifying factors are:

1. *Molecular weight and distribution.* It is usual for these factors to be contrived during the manufacturing processes, but under production conditions it is not easy to ensure constant results. Some studies have shown that a few percent of a low molecular weight material can halve the fatigue life of a specimen or component.
2. *Degree of cross-linking.* This usually has a marked effect on the modulus and other physical properties, but the degree of cross-linking may change with time, usually in the direction of further cross-linking. These progressive changes may be a natural tendency within the material, but cross-linking can be affected by external factors such as work hardening, work softening and the environment, i.e. absorption or ultra-violet radiation.
3. *Composition.* All commercial plastics contain additives such as plasticisers, extenders, pigments, fillers and terminators. These can greatly modify the performance of the material. Some inclusions actually form the seat of a fracture whilst others act to reinforce the material or to blunt cracks.
4. *Material orientation.* The manner in which the material is processed will not only affect the distribution of the material but will also create distinctive flow lines and stress patterns (see point 5). Furthermore the materials may be co-polymers or composites, in which case the distribution can be even more dramatic.
5. *Thermal history.* This effect extends from early in the manufacture of the basic material through to the finished component. Such factors as flow, shrinkage, internal stresses and

surface condition have to be considered at each stage and some interstage thermal treatment may be necessary. Temperature is another of the vital factors found in the working environment cited above in point 2.

This generalisation of materials has no absolute validity. Indeed, in the literature there is, in most cases, strong evidence refuting or modifying the claims made in other papers or caveats stating that the results have no universal applicability. For instance structural differences within the same basic material can cause crack growth rates to be modified by four orders of magnitude! It is regrettably the case that comparatively little work has been done on fatigue in polymers. This lack of fundamental work on fatigue has restricted progress greatly. The shear diversity of properties within each polymer category, coupled with the lack of formalised data available to the designer, has resulted in the low usage of plastic for engineering applications. Since disenchantment by the user is one of the major driving forces in the search for failure criteria, it follows that, without extensive application, working and design data will be limited and vice versa. The above are some of the factors within the polymer, but there are others that directly relate to the manner in which the plastic is used and where it is used.

4. FACTORS RELATING TO THE USE OF THE MATERIAL

It is impossible to consider all the possible applications, but there are two generalised areas that must receive particular attention in any design:

1. The test/working environment.
2. The frequency, amplitude and nature of the applied load.

In many cases a principal factor in the choice of a plastic is its corrosion resistance. However, plastics are themselves vulnerable to some environments such as solvents, certain detergents, acids, alkalis, ultra-violet radiation, mineral oils (natural rubbers), and even air and water in certain instances. All of these can have a marked effect on the performance of the polymer particularly under fatigue loading.

It has already been stated that the viscoelastic behaviour of the material is the most crucial factor in restricting fatigue studies. Even at relatively low loads there is significant hysteresis in the load cycle and since this represents work, which appears as heat, there is a problem of heat removal. This is exacerbated by the low thermal conductivity of most plastics. The results are that thermal gradients occur within the material which may be increased if the heat is forcibly removed.

The frequency and rest periods of a test are quite crucial to creep or fatigue life. Viscoelasticity implies a spring–dashpot model for the material, which by definition is time-dependent. Thus there will be a delay in the response, an inherent error and a marked difference between the applied stress and the resultant strain. There is a general acceptance that this difference in response is only reduced significantly if that particular state of stress or strain is maintained for a considerable period of time (of the order of 400 s). Thus there is always the problem in dynamic testing of knowing the precise nature of the applied load. There is also a further complication with dynamic testing of deciding the appropriate rate of load application. In general, most materials respond unfavourably to the sudden imposition of load. The severity of this load regime will depend on the frequency, extent and profile of the load. The profile of the signal may be random (white noise), but quite often it is sinusoidal, saw-tooth or square waves. Square waves are clearly the most severe for any given frequency or amplitude. In practice, therefore, there is no absolute control over the precise stress/strain conditions within the material, especially if there is strain softening or strain hardening.

5. LOADING STRATEGY

It is virtually certain that whatever the loading pattern necessary for the test, the response of the machine will not match the condition of the actual load in any direct way and the loading axes will have to be modified to obtain that match. In addition, some attempt must be made to accelerate the test programme. However, it is not sufficient to accept so-called 'stable failure' (that is any test where the specimen does not fail uncontrollably in a thermally dominated way) since this may be totally unrepresentative of the actual service conditions. To avoid this thermal problem, an alternative is to enhance the stress or strain and to markedly reduce the frequency. The advantages of this approach are:

1. It still may allow frequencies to be used which are considerably above those in the loading regime thereby gaining time.
2. The temperature will have a chance to stabilise to a level similar to that expected in the actual application and thermal gradients within the specimen may be less severe.
3. The stress/strain relationship will be more closely related and observable.
4. Creep effects can be observed in terms of cumulative damage and the converse of allowing creep recovery to operate. Any creep recovery that is possible will reduce the 'latching' effect of progressively hysteresis loading.

As previously stated, the environment can also be crucial to a component in service and the effects of such factors as ultra-violet light, water, solvents and temperature would have to be assessed in relation to the specific conditions. No test facility would be complete unless there was some provision to control the environment. Consequently, the design must include sufficient working space for modifying the test environment, if only locally, in a similar manner to that found in (say) environmental cabinets of standard tensile testing machines.

The control can only be through sensors that are stress- or strain-sensitive (there are other techniques such as infra-red and sound emission, but these are very much in the developmental stage) and there are three basic systems:

1. Stress-dependent.
2. Strain-dependent.
3. Stress or strain above the mean level of stress or strain.

These can lead to quite different end results. In strain-dependent tests there is usually a marked change in stress, since there is a tendency for the material to either strain-soften or strain-harden. Strain softening is very common and means that failure may not necessarily proceed. Stress-dependent tests can, under similar circumstances, progressively increase the strain to eventual failure. Mean stress can be present in a component for several reasons:

supporting a load, or 'bolting up' load, inherent stresses or strains, temperature distortion, etc. These are not necessarily detrimental to performance, but, on the contrary, may assist if, say, these stresses are compressive when the external load is tensile. There are other more subtle benefits that have been suggested such as crack blunting and energy absorption during molecular reorientation. Overstrain can also have the beneficial effect of slowing a fatigue crack, conflicting with some of the cumulative laws used in metals.[1]

6. INSTRUMENTATION AND CONTROL

The effectiveness of the whole experiment is controlled by the efficiency of the measuring system since the control functions of the machine are reliant on an accurate 'feedback' of information and the subsequent analysis of material properties. In general, conventional extensometry that functions by physical attachment to the specimen does not perform well under dynamic conditions. Several reason for this exist:

1. Delicate mechanisms do not sustain dynamic loading very well and are prone to fatigue failure themselves.
2. The mechanisms rely on high velocity ratios for their amplification, which, because of stiffness considerations in the component parts, lead to unfavourable disposition of the mass centres of the linkages, etc. (This is another factor in favour of low-frequency testing).
3. Because of the need to ensure firm attachment to the surface, the gauge causes indentations in the specimen. This can lead to premature failure at that point. This is especially true if there are very high strains within the specimen.
4. The design is usually for uniaxial application. It is these limitations in the meaurement of strain that have resulted in a great many tests in the field of polymers controlled by stress levels which can be remotely sensed away from the actual specimen without undue loss of precision. These comments relate to uniaxial tests and the translation from the uniaxial to multiaxial mode has proven to be difficult even for metals which, in general, are easier to gauge because of the lower strains involved.

The above represents a formidable list of factors that discourage any work on fatigue behaviour of polymeric materials, but if the material is to be fully exploited then the performance characteristics must be further understood and the likelihood of a failure of a particular material with a given configuration estimated.

7. TEST FACILITY

There are many multiaxial fatigue machines reported in the literature, a large proportion of which are biaxial (cruciform) systems. Very few are designed for universal application and most require specimens that have highly specialised geometry for ease of mathematical analysis or to force the failure into a prescribed zone. Much has been gained from these tests, but cross-correlation between one system and another is extraordinarily difficult since each test yields data unique to that configuration. To obviate some of this limitation the requirements of a more generalised system are discussed and, in particular, the more unusual design aspects of a family of machines that have been especially developed for this purpose. Since an earlier variation of the basic design has been well proven on metals[2,3] and because the testing of plastics involves many additional problems that are even more difficult to resolve, it is only those aspects that are unique to plastics that will be considered in detail.

8. ESSENTIAL DESIGN FEATURES

As indicated above, there are many design possibilities, but a basic requirement of any facility is for two surfaces that move toward and away from each other (push–pull), and that those surfaces should rotate with respect to each other (right- and left-hand torsion). A practical means of obtaining a triaxial state is to provide a high-pressure fluid to a hollow component (hoop stress), see Fig. 1. For adequate simulation these systems should operate in-phase or out-of-phase with each other, either stress- or strain-dependent. They should have common specimen geometry and chucking and either prescribed waveforms or a facility to accept an external random (non-specific load) signal. Whilst a multiaxial machine has to be

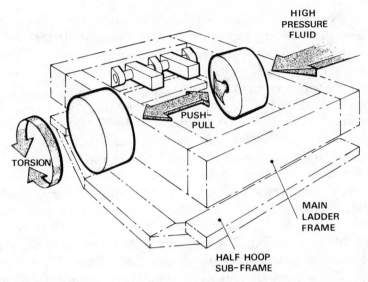

Fig. 1. Practical system for triaxial fatigue loading with indicated support frame.

designed as a complete entity, for clarity the basic elements of the machine will be considered separately as far as is possible. Since the potential loads can be applied in any combination, the basic design philosophy of the machine will be totally stiff,[4] that is, the structure and bearings will resist any combination of load without any significant deflection.

9. MACHINE STRUCTURE

The mainframe of the machines was designed to be a ladder-like box with a push–pull load acting down the centre, and the sub-frame as a half-hooped cage to resist the torsional load, as outlined in Fig. 1. All the structures were fabricated from heavy steel section—very much over-designed by conventional standards, but thereby completely preventing any spurious deflections.

9.1. Torsion systems

When a specimen fails under compression, its basic instability causes it to buckle, thus imposing a bending moment in addition to the axial load. In torsional loading of fibre-wound pipes the failure mode becomes asymmetric because of delamination along certain of the fibres and this can also apply a bending moment. If the torsional surfaces are to remain perfectly normal to each other, then the bearings that allow the rotation must be housed in such a way as to resist this bending moment and any axial loading. The other surface must also be stable against these extraneous forces, but must resist the torsional load. Torsional loading is usually accompanied by a change in length along the axis and therefore the restraint against torsion must allow translation. The stiffness of the rotational head is obtained by having either tapered-roller or angular-contact bearings, spaced apart in a very solid head, thus acting as two very substantial rungs of the 'ladder-frame'. Since the bearings have to withstand the compressive load as well as the tensile load, they have to be given considerable pre-tension. These bearings are also overdesigned, by normal codes of practice, for two reasons: they have to provide this 'totally stiff' base, and there has to be sufficiently large inner diameter to permit a hollow shaft for the access of the internal pressurisation fluid. This arrangement offers a negligibly small frictional resistance. The drive to the torsional system depends on the load actuator which can be servo-hydraulic or stepper motor/gearbox. There can be backlash problems in both systems which can be minimised by optimising the gearbox assembly against 'backlash' or having a rubber (toothed) belt intermediate drive in the case of the stepper system. The servo-hydraulic actuator can be arranged to drive on an opposed cylinder arrangement, but this then requires a unit twice the designated load capacity, but does eliminate 'back-lash' completely. Stepper motors give very precise control but require considerable adaptation for total computer control and are very noisy in operation.

9.2. Axial loading system

The axial loading mechanism has to incorporate three major features. Firstly it must be braced against the torsional load, secondly it must be constrained on-line to prevent any misalignment due to buckling, and thirdly it must contain a load cell for stress in torsion as well as tension and compression. It also has to contain the mechanism for the internal pressure compensation (see Section 9.3). Robertson and Newport[5] showed that there can only be axial stability if the loading mechanism is rigidly constrained on a line by a series of bearings within a ladder-framed structure and this has been a guiding principle in the design, but equally one end of the specimen has to both rotate and translate as well as be held on the load axis. To meet this requirement the chuck was supported in a large needle roller mounted in a stabiliser frame. The needle roller was both large and had a long needle length, this gave virtually the same frictional resistance but with sufficient axial constraint to resist buckling. It is convenient, although not necessary, to have this chuck stabiliser frame running in the same linear bearing as that used for the torsional resistance arm. The torsional load has to be resisted at this end and it is convenient to resist this load at a point behind the stress transducer but before the axial load piston. Again translation has to be permitted but all other movement resisted. Thus this interface of the loads (thrust block) has arms which are constrained by similar linear bearings to those used in the stabiliser. The bearing tracks are widely spaced to increase the moment arm and reduce the side load on the linear bearing. These bearings are crucial to the design since they have to sustain the very high torsional loads but offer very low resistance to axial motion and constrain the whole assembly on the load axis. The actual load is applied through a double-rodded cylinder directly coupled into the thrust block. This cylinder is actuated via a servo-control valve.

9.3. Internal pressurisation

This is the most complex of the three systems and has a considerable effect on the design of the other axes. Figure 2 shows a typical chucking arrangement, the complexity is primarily due to the presence of pressurising and environmental liquids and the separation and sealing of them. The design is constrained by three main factors: the fluids, the pressure, and the frequency and nature of the load.

Whatever the choice of fluid in contact with the walls of the specimen there has to be a separation of the pressurising fluid from the hydraulic fluid which is usually filtered down to a level of 2 μm. In any case most plastics are affected in some way by oils and their exclusion from the specimen must be possible. There has to be an intermediate phase which protects the integrity of the hydraulic oil and the environmental fluid. If the environmental fluid is hostile to the intermediate stage (i.e. if it is water or an acid), then the

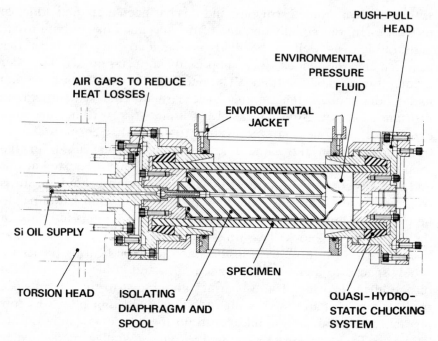

Fig. 2. Specimen mounting system with external water-heating jacket. Specimen is a piece of standard gas pipe (90 mm OD) cropped to length.

intermediate fluid would have to be an oil. Except in work on stress corrosion, most metals function well with oil as a second-stage fluid. If the environmental fluid is particularly hostile, then the volume is kept to a minimum. This is achieved by introducing a highly flexible membrane within the specimen and the overall volume is reduced by putting the membrane around a spool.[6] The design shown in Fig. 2 is a system where the pressurising fluid (second stage) is silicone oil and the environmental fluid is water. Apart from the obvious problems of sealing a system under multiaxial stress there are other problems such as loading the component into the machine, markedly different response characteristics, extensometry, safety, spillage, failure detection, etc. A particular problem is the induced axial load that results from internal pressurisation. This constant ratio of axial load to hoop stress has been obviated by the introduction of a supplementary piston of the same diameter as the inner bore of the specimen but remote from it; this has enabled this axial component of the load to be totally flexible.[6]

9.4. Control

Simulation requires that the loading pattern closely resembles the actual applied load and whilst there is the possibility of load enhancement the technique is essentially a unique method because of the time constraint. Consequently the control is of extreme importance, especially since, as indicated above, the response characteristics of the various modes are quite different and will vary with each specimen configuration. There are a number of factors that affect the performance of the system, viz:

1. The input signal quality.
2. The control and response of the servo systems.
3. The basic integrity of the machine.
4. The extensometry.
5. The specimen preparation, geometry and gripping.

Whilst the hysteretic heating of the materials prevents high-frequency methods, it means equally that the loading within the simulation has to be comparatively slow, which is an advantage from the control aspects of the machine. The load programme is derived from the actual component and will have been taken from that part of the component that is most adversely loaded. This load has to be translated into the gauge length of the test specimen by interpolation between the actual load and a combination of the basic loading regimes available on the machine. This requires considerable manipulation that is only possible with a sophisticated computer system. The system requires a high degree of interaction since the loading can be random and varying over a long period, especially if the test is not to become quasi-cyclic. This costly requirement has prevented this degree of refinement on these machines to date. However, there are many real-life situations that are cyclic and under these conditions a comparative load can be 'tuned' into the machine. This is possible because each axis of the machine is equipped with pseudo derivative feedback (PDF) control. These axes can be controlled within 2% of each other. However, because of the comparatively slow response of the internal pressurisation system, the other systems are controlled

from the output transducer of this system (cascaded) when synchron-
isation is required with the internal pressurisation system.

The basic integrity of the machine is essential since the loading
direction is non-specific and any untoward deflections of the machine
would affect the extensometry which has to be three-dimensional.
Items 4 and 5 above present major problems and there is a need for
much research in these areas. The three-dimensional nature of this
work merely makes this already difficult problem worse. For example,
conventional extensometry relies on physical attachment to the specimen. This is usually in the form of lightly applied knife edges or
points. This type of attachment is quite prone to induce premature
failure that is directly attributable to the actual contact, but when the
material beneath the contact point is undergoing multiaxial strain the
problems increase. There is an increase in the application of remote
sensing devices to detect factors such as light, induction, infra-red
and acoustic emissions, but these are usually more sophisticated and
temperamental, and it is more difficult to obtain continuous signals
for control purposes. A system of air gauging was developed for this
work and this solved many of the problems associated with the other
systems, but it was found that the low-pressure air impinging on the
surface was affecting the material.[4]

The problems of specimen preparation[7] and geometry are legion
and because of the apparent insolubility of these difficulties each
experimenter makes some accommodation adjudged suitable for the
series of tests, and in very few cases does the actual method seem
adequate. In the case of simulation the problem is extremely difficult
to resolve because not only has the grade and composition of the
material to be truly representative, but it should also match the
production condition, which if it is (say) a particular section of an
injection moulding is almost impossible to create in the gauge length
of a specimen. Consequently, because the test is essentially unique, the
preparation of the specimen, especially the thermal history, is abso-
lutely vital.

The problem of gripping the specimen is another very important
factor in the testing of plastics. Almost every experiment is modified
by the 'end effects' of the specimen grips or the way the load is
supported. The three-dimensional nature of the intended loading of
these machines presents particular problems with respect to 'end
effects'. A quasi-hydrostatic system has been developed[4] with rubber

acting as the fluid. This has proven to be most successful and has worked in every application tried so far, even with the very high loads associated with large glass-wound epoxy pipes.

10. CONCLUSION

There are numerous limitations to the dynamic testing of commercial plastics, especially as the collection of data by the traditional methods associated with metals is impractical. However, if plastics are to be used in engineering applications, a knowledge of their dynamic performance is essential. An intermediate system has been developed which allows a direct measure or model of the load regime to be applied to the gauge length of a test specimen. Whilst this system falls short of full-scale testing it represents a considerable advance over techniques which rely on correlated uniaxial data.

The methods described enable fatigue tests to be carried out within the standards usually associated with conventional test machines. Since there is common geometry and gripping for all loading modes any combination of load is possible, thereby enabling a comprehensive programme to evolve from simple tests such as conventional tensile, compression or torsion tests. This versatility also enables comparative tests to be made with existing- data.

ACKNOWLEDGEMENTS

It is impossible to mention individually the many people who have assisted in the many years of development, but their contribution is recognised with gratitude. However, the continued assistance from S.E.R.C. over recent years has been of especial help and for this I am extremely grateful.

REFERENCES

1. Turner, S. (1966). The strain response of plastics to complex stress histories, *Polymer Eng. Sci.*, (October) 306.

2. Liddle, M. and Miller, K. J. (1978). Multiaxial High Strain Fatigue, Proceedings 3rd International Conference on Fracture 1978 Munich.
3. Brown, M. W. (1975). *High Temperature Multiaxial Fatigue*, PhD Thesis, University of Cambridge.
4. Lawrence, C. C. (1980). Triaxial fatigue testing machine for polymeric materials, *J. Strain Analysis*, **15**(3) 159.
5. Robertson, A. and Newport, A. V. (1927). *Report on the Drop of Stress at Yield at Armco Iron*, Report Memoranda Aero, M56 Committee, London No. 1161, p. 1.
6. Lawrence, C. C. to be published, ASTM Proceedings, 1983.
7. Lawrence, C. C. (1979). High temperature post manufacture heat treatment of thermoplastics, *British Polymer J.*, **10**, 93–7.

Polymer Testing **4** (1984) 289–305

Fatigue Life Determinations of Rubber Springs

A. Stevenson

The Malaysian Rubber Producers' Research Association, Tun Abdul Razak Laboratory, Brickendonbury, Hertfordshire SG13 8NL, UK

SUMMARY

A method for the determination of fatigue lives of rubber springs is described which is based on a fracture mechanics approach. Fatigue experiments have been performed on model components consisting of solid rubber cylinders bonded between metal endpieces of the same diameter. Crack growth in uniaxial compression followed an approximately parabolic locus. Tearing energies were determined both experimentally and theoretically. The results could be plotted on a unique curve of tearing energy, T, versus crack growth rate, dc/dN, as could results from tests on thin strips of rubber fatigued in simple extension. Application of the approach is described to two case histories. The first is the design of a large spherical rubber joint to support an offshore oil-production platform, and the second is the fatigue analysis of a rubber antivibration mounting for a rail car after 2·5 years service.

NOTATION

A Crack area (mm²).
c Crack length (mm).
D Testpiece diameter (mm).
e Strain.
e_c Compression strain.

Polymer Testing 0142-9418/84/$03·00 © Elsevier Applied Science Publishers Ltd, England, 1984. Printed in Northern Ireland

e_H Horizontal surface strain.
e_v Vertical surface strain.
E_c Compression modulus (GPa).
G Shear modulus (MPa).
h Testpiece thickness (mm).
k Geometrical factor.
l Constant displacement suffix.
N Number of cycles.
S Shape factor = loaded area/force-free area.
T Tearing energy (kJ m^{-2}).
U Stored strain energy (J).
W Strain energy density.

1.   INTRODUCTION

Rubber springs can be designed to act as flexible bearings or mounts to support a wide range of engineering structures—from automobile engines to buildings or even offshore oil rigs. Such rubber bearings usually consist of one or more elastomer layers bonded to reinforcing metal plates. Determination of the fatigue life of such a unit requires first the clear definition of a criterion for failure. This will relate to the performance of the component and may be expressed in the form of a percentage-acceptable change in some vital engineering property (such as stiffness), or directly in terms of the growth of cracks to some acceptable limit. Fracture mechanics techniques can be applied to determine the rate of crack growth (dc/dN) in the elastomer for the given conditions of dynamic loading. The loading conditions of an actual component may not always remain the same, so that, in practice, spectra of loads (and motions) need to be considered for their combined contribution to fatigue damage.

There are few published references to the fatigue of rubber in compression. Early investigations by Cadwell et al.[1] suggested that failure could occur when rubber blocks were subjected to repeated compression. Hirst[2] has reported that rubber railway buffers, in the form of bonded cylindrical blocks, had chunks of rubber removed from the exposed edges similar in form to that depicted in Fig. 1. Lindley and Stevenson[3] have reported a preliminary investigation of fatigue crack growth in non-bonded rubber blocks compressed be-

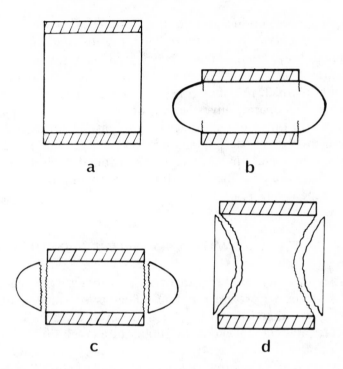

Fig. 1. Stages of crack growth in a bonded rubber compression spring. (a) Initial—unstrained; (b) strained—cracks initiate at bond edges; (c) strained—ring cracks melt to separate a collar of rubber; (d) unstrained—showing parabolic locus of crack growth.

tween rough (high-friction) platens. The present paper reports on the extension of this investigation to rubber discs of various sizes and shape factors bonded to metal endpieces. A fracture mechanics analysis is used, determining tearing energy, T (equivalent to strain energy release rate) and crack growth rate, dc/dN.

The relationship between tearing energy and crack growth rate is compared with that obtained from more convenient fatigue crack growth tests in simple extension. A simple theoretical description of tearing energy under compression is also provided.

The methods will be illustrated by detailed reference to an experimental study of fatigue crack growth in model rubber spring components consisting of bonded rubber cylinders subjected to cyclic compression. Rubber springs are most frequently used in compression in engineering applications and so this represents perhaps the most important case. In these experiments the tearing energy T was determined both from simple theory and also from experiments. Crack growth rates were in this case found to be independent of crack length, which provides considerable simplification and reduces the need for a separate investigation of crack initiation conditions.

The application of the approach to two case histories is discussed. The first involves a design analysis of a large laminated natural rubber bearing for use in an offshore oil-production platform, and the second consists of a failure analysis of an antivibration pad used in a rail-car suspension.

2.   CRACK GROWTH AND FATIGUE IN RUBBER

Two types of crack growth can occur in rubber. One, chemical in nature, is caused by ozone, but its effect is so small in the work described here that it need not be considered further. The other is small-scale physical tearing of the rubber at a crack tip. This can lead to very rapid crack propagation when the energy available for such tearing (the 'tearing energy') exceeds a critical value T_c, (usually of the order of $10\,\mathrm{kJ\,m^{-2}}$. The tearing energy is formally equivalent to the 'strain energy release rate' referred to in fracture mechanics studies of plastics and metals.

In a crystallising rubber (such as natural rubber, NR), held at constant deformation, strain-induced crystallisation (highly hysteretic in nature) of the rubber at a crack tip prevents crack growth below T_c.

For repeated deformations the amount of crack growth which occurs in a deformation cycle depends on the maximum tearing energy, T, during that cycle. For a given T the maximum amount of growth occurs when the rubber is completely relaxed during the cycle. For such relaxing conditions and $T < T_c$

$$\frac{\mathrm{d}c}{\mathrm{d}N} = f(T) \tag{1}$$

where c is the crack length at number of cycles N. There is a minimum value of T (T_0) below which no mechanical fatigue crack growth will occur.[4] This value is, however, much lower than the values observed in the present tests and so need not concern us here. The relationship (eqn. (1)) betwen dc/dN and T is thought to be a material property, independent of testpiece geometry. This has been confirmed for several types of tensile testpiece[4] and for bonded shear testpieces.[5] The main aim of the present investigation was to study the validity of this approach for bonded rubber units subjected to applied compression.

3. EXPERIMENTAL

The compression testpieces consisted of rubber cylinders bonded to metal endpieces of equal diameter. The testpiece shape factor S was varied from 0·25 to 6·5 and rubber layer thickness varied from 2 to 50 mm. The rubber formulation in parts by weight was natural rubber (SMR CV60):100 and dicumyl peroxide:1. The compounded rubber was vulcanised in metal moulds for 60 min at 160 °C. This formulation gave a translucent rubber which enabled the progress of any cracking into the rubber to be seen with the aid of back illumination. The rubber–metal bonds were formed during vulcanisation in the moulds using the bonding system Chemlok 220/205 supplied by Hughson Chemicals, USA. These testpieces were model components—similar in geometry to many actual rubber springs.

Relaxing fatigue tests were carried out on a modified 12·5 ton Hedley die-stamping press at a frequency of 2 Hz. On each cycle the testpieces were compressed from the unstrained state to a chosen compressed thickness to give a constant amplitude in the range 15–50%. The testpieces were removed at intervals for inspection. There was no evidence of either internal cracking or substantial heat build-up—all cracks initiated at outer rubber surfaces.

In addition to the fatigue tests in compression, a set of fatigue tests was carried out in simple extension. Strip testpieces were cut from rubber sheets moulded from the same batch of rubber compound used for the compression testpieces. Relaxing fatigue tests were again carried out at a frequency of 2 Hz.

4. DETERMINATION OF TEARING ENERGY

The tearing energies T for these rubber testpieces[6,7] were determined from

$$T = -(\partial U/\partial A)_l \tag{2}$$

where U is the total strain energy in a testpiece containing a crack of area A (area of one surface of the crack). The suffix 1 denotes that differentiation is carried out at constant displacement.

For a tensile strip with a single edge crack of length c, it has been shown[6] that

$$T = 2kWc \qquad (3)$$

where W is the strain energy density in simple extension at strain e, and k is a strain-dependent geometrical factor which has been determined experimentally[8] and by finite element analysis.[7]

For a bonded rubber block in compression the strain energy density W will depend on the shape factor. The shape factor S of a bonded rubber block is defined as the ratio of loaded (i.e. cross-sectional) area to the total area of force-free surface. For a disc of diameter D and thickness h

$$S = D/4h \qquad (4)$$

Rubber has a high bulk modulus (~2 GPa) relative to its Youngs modulus E (~0·002 GPa) so that most deformations occur with negligible volume change. When a cylinder bonded between rigid end plates is compressed, the rubber bulges to maintain constant volume. For a small compressive strain the profile of the bulge is approximately parabolic. The change of slope of the parabola from the unstrained state is the shear strain. Shear strains occur throughout the cylinder and their elastic energy is additional to that of the normal strains, giving rise to an effective compression modulus E_c greater than the Youngs modulus

$$E_c = 3G(1 + 2S^2) \qquad (5)$$

where G is the average small strain shear modulus. To a first approximation, the stored strain energy density W may be expressed as

$$W = \tfrac{1}{2}E_c e_c^2 \qquad (6)$$

where e_c is the compression strain.

If the tearing energy is assumed[9] to scale with the unstrained distance between bulge centre and bond edge, $h/2$, then one may write

$$T = \tfrac{1}{2}Wh = \tfrac{3}{4}G(1 + S^2)e_c^2 h \qquad (7)$$

Fatigue crack growth experiments were carried out for a range of different shape factors S and thicknesses h, to test eqn. (7).

The tearing energy T was also determined experimentally for each compression testpiece. This involved removing each testpiece from the fatigue machine at several intermediate stages of crack growth, and obtaining a force–deflection curve in compression on an Instron testing machine. The area under the force–deflection curve was measured, thus determining the stored strain energy U. The difference between this and the stored energy just before crack initiation gave the total energy loss ΔU due to the total amount of crack growth. At each stage the total crack area was measured.

Values of ΔU were plotted against ΔA, and (in accordance with eqn. (2)) the tearing energy was given by the slope $\Delta U/\Delta A$. In almost all cases the plots of ΔU versus ΔA were linear through the origin as shown in Fig. 2. Thus crack growth in compression appears to be characterised by a constant value of T. The experimental values for T obtained in this way turned out to be in a good agreement with the theoretical values calculated from eqn. (7).

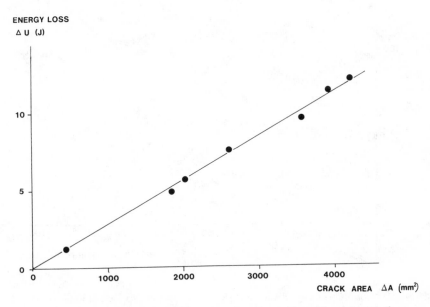

Fig. 2. Typical measured energy loss versus area of crack growth.

5. SURFACE STRESSES AND STRAINS—CRACK INITIATION

Cracks open up under the action of tensile stresses and, at a force-free surface, are most likely to initiate in the plane perpendicular to the maximum tensile stress at the surface.

The rubber at the surface of the bulge is in a state of two-dimensional plane stress. The principal stresses are the hoop (or horizontal) stress σ_H and the surface stress normal to σ_H, which for convenience is called the vertical stress σ_V. The principal strains corresponding to σ_V and σ_H are e_V and e_H, respectively.

The surface strains were measured at various locations on test-pieces of shape factors 0·25, 0·5 and 1. Following Gent,[10] small circles of known diameter were made on the cylinder at various compressions e_c and the diameters of the resulting ellipses measured in the unstrained state to determine e_V and e_H. Average values of e_V and e_H are shown plotted against applied compression in Fig. 3 for

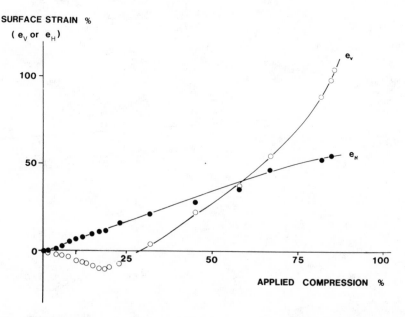

Fig. 3. Typical measured tensile surface strain versus applied compression. O, Vertical component; ●, horizontal component.

one typical case. These results suggest which types of crack are most likely to initiate.

Vertical strain e_v, tensile positive, was generally greater than e_H and greater near the bond than at the bulge centre. This is consistent with the observation that circumferential cracks initiated most commonly at the bond edges. However, at low values of e_c (below about 25%) e_H became greater than e_v. The exact position of the cross-over varied with shape factor but this predicts radial cracking at low strain levels. At low amplitudes of compression strain e_c there are experimental difficulties in performing fatigue tests due to the very large numbers of cycles required (10^7–10^8). However, two such tests were carried out and radial fatigue cracks were indeed observed, as predicted by the analysis of surface strains.

6. FATIGUE CRACK GROWTH—RESULTS

It was an interesting feature of crack growth in compression that both the tearing energy, T, and crack growth rate, dc/dN, were approximately constant throughout each test. This is not so in simple extension, where T, (and thus also dc/dN) increased strongly with crack length c. Values of T from compression tests are plotted against dc/dN in Fig. 4 as circles. Also shown in Fig. 4 is the envelope of T versus dc/dN values obtained from the fatigue of simple extension testpieces of the same rubber vulcanisate. These two sets of experimental observations were in good agreement confirming the view that the relation (defined by eqn. (1)) is a material constant completely independent of testpiece geometry.

The values of tearing energy were both determined experimentally and calculated (using eqn. (7)) for the compression testpieces. The calculated values of T are shown in Fig. 4 as open circles and the experimental values as closed circles. Agreement was obtained between the two methods. This suggests that eqn. (7) is valid at least over shape factors between 0·25 and 6·5 and for compression strains between 15 and 50%.

Crack growth in bonded rubber units under compression was restricted to the vicinity of an outer 'bulge region' where local shear strains are appreciable. In each case a central core was left without visible crack growth. The size of this core relative to the testpiece

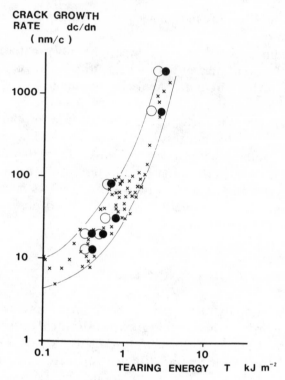

Fig. 4. Relationship between tearing energy and crack growth rate. ×, Simple extension; ●, compression (measured *T*); ○, compression (calculated *T*).

diameter decreased with increasing test strain and testpiece shape factor. Figure 5 shows a typical testpiece after fatigue crack growth had completely removed the outer rubber collar to expose the central parabolic core.

7. APPLICATION TO THE DESIGN OF A RUBBER ARTICULATED JOINT FOR AN OFFSHORE OIL-PRODUCTION PLATFORM

The deep-water gravity tower[11] has been designed by a major constructor of offshore oil platforms (C. G. Doris) to provide a support for drilling and production of hydrocarbons in great water depths (see Fig. 6).

Fig. 5. Photograph of fatigued testpiece showing parabolic locus of failure. The rubber collar (to the right) separated by fatigue crack growth.

The column is composed of a concrete floater supporting the deck and a steel jacket terminated at the lower part by a ballast chamber. This unit has a global downwards apparent weight at the level of the articulated joint, located between the column and the fixed steel base, piled to the sea bed. Considering that it is kept permanently in compression, an original type of articulated joint has been developed based on laminated rubber pads made from alternate layers of rubber and metallic shims bonded together during vulcanisation.

The ball joint for the deep-water gravity tower consists of a series of laminated rubber pads inserted between two hemispherical steel shells, as shown in Fig. 7.

A central circular pad, about 1·20 m in diameter, is surrounded by two rows of eight trapezoidal pads. Each pad is composed of a sandwich of layers of rubber, 10–15 mm thick, and of curved metallic plates, about 6 mm thick. The two extreme plates are 40 mm thick. The pads are securely fastened to the inner shell which is connected

A. Stevenson

Fig. 6. Artists impression of the deep-water gravity tower. The rubber ball joint is located at the base of the tower.

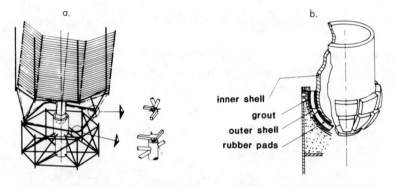

inner shell
grout
outer shell
rubber pads

Fig. 7. Rubber articulation system for the deep-water gravity tower. (a) Principle of articulated joint and torsion frame; (b) detail of articulated joint.

to the central vertical member of the steel jacket, and to the outer shell which is located in the base central member. When a rotation is imposed to the articulation by the heel of the tower, the inner shell rotates and imposes a tangential distortion to the pads which develop only negligible resisting forces, thanks to their low shear stiffness.

Nevertheless, the elastomeric segments have a very high stiffness in the axial direction, so that variations of vertical reaction, due to live loads on deck or to effects of environmental loads, result in very little vertical displacement.

The fracture mechanics approach described in this paper was used to estimate the life of the laminated rubber segments of the ball joint. A spectrum of loads and motions for the whole tower was determined with the help of a specially developed computer program and checked against extensive basin model tests corresponding to both regular and irregular waves. The tower loads and motions in the spectrum were then analysed into elastomer strains in the ball joint. The life estimates for the ball joint were then made with the help of the characteristic tearing energy–crack growth rate relationship— similar to Fig. 4.

The results of this analysis could then be used to influence improvements in the design. The optimisation of the design for fatigue life was thus placed on a rational basis by means of the fracture mechanics approach.

8. APPLICATION TO A RAILCAR SUSPENSION UNIT

Figure 8 shows two rubber pads bonded to steel mounting plates which form part of the suspension system of a railcar designed and built by Metro-Canada. After 2·5 years service Metro-Canada en-

Fig. 8. Rubber bolster unit for railcar suspension. The rubber is bonded between the outer steel plates.

gineers observed some cracks and surface discoloration and, although the performance of the spring was still satisfactory, raised questions concerning its likely further life. The rubber layer was debonded from the metal plates and sectioned to reveal the depth of the cracks visible on the surface. A typical section is shown in Fig. 9.

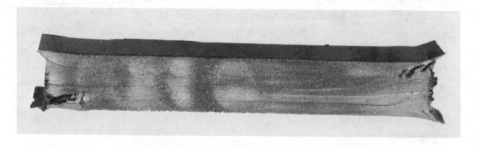

Fig. 9. Section through rubber layer removed from the bolster unit of Fig. 8.

A large number of very fine surface cracks were revealed which were only 0·2–0·4 mm in depth. These were associated with the discoloured regions and attributed to a combination of photo-oxidation and ozone attack. This form of crack growth is essentially only a surface effect and is unlikely to have any significant effect on the spring rate of the unit. In addition, larger cracks were observed at the bond edges apparently growing into the rubber at a shallow angle. The angle of crack growth is broadly that expected under conditions of combined dynamic shear and compression. At one pair of diagonally opposite corners, crack growth to a depth of 5–6 mm was recorded and for the other pair it was to a depth of 1–2 mm. The locations of the larger cracks corresponded to where the local shear strain due to compression was in the same direction as the overall shear—and the two contributions added. In the opposing corners the contributions were of opposite sign.

The exact service loading conditions are difficult to determine since they depend on, for example, the height of unevennesses in the rail track, loading variations during service from day to day, number of stops and starts, etc. However, it is known that on a test track of 2 km, 10^5 dynamic load cycles were recorded of amplitude between 3 and 23 kN, in every case superimposed on a static load of 22 kN. Calculations of tearing energy by means of eqn. (7) indicate that the minimum static tearing energy is 0·4 kJ m^{-2} and the maximum tearing energy is 1·7 kJ m^{-2}. The ratio T_{min}/T_{max} therefore varies between 0·71 and 0·23. Taking into account the effect of the non-zero minimum, the range of crack growth rates this suggests is $<10^{-4}$–$>0·5$ nm/cycle. It was known that the total distance was 10^5 km. This suggests a total of 5×10^9 events and crack growth between 0·5 and 250 mm. Although no load spectrum is available it is likely that the largest number of cycles is at the low severity end so total crack growth of 1–10 mm would appear to be the likely prediction from the theory. This is consistent with the observed crack lengths of 5–6 mm in one case, and 1–2 mm in the other.

The value of the above (somewhat approximate) analysis is that it provides some understanding of the causes of crack growth. It provides the reassurance that crack growth is stable—so that a further 2·5 years service would appear feasible before any dramatic change in the spring rate characteristics occurred. It also gave a rational basis for Metro-Canada to formulate an improved 'second generation' design for even longer life.

9. CONCLUSIONS

A fracture mechanics approach has been successfully applied to analyse crack growth behaviour in bonded rubber cylinders under dynamic compression. The relationship between tearing energy and crack growth rate was the same as that obtained from fatigue crack growth testing with tensile strips of the same rubber. This is conclusive evidence that the relationship is a fundamental material property. A mathematical expression was proposed for the tearing energy and shown to be valid for a range of shape factors and compressive strains. Fatigue crack growth occurs in compression at a constant rate and tearing energy, and is restricted to the outer regions of the test pieces with high local shear strains. The crack loci can be predicted from an analysis of the surface strains in compressed uncracked testpieces.

The two applications discussed illustrate the practical value of the tearing energy approach in providing a rational basis for evaluating the design of new components and for the failure analysis of components in service.

ACKNOWLEDGEMENTS

This work forms a part of the research programme of the Malaysian Rubber Producers' Research Association. The author would like to thank C. G. Doris Ltd and Metro-Canada Ltd for permission to describe the two applications.

REFERENCES

1. Cadwell, S. M., Merrill, R. A., Sloman, C. M. and Yost, F. L. (1940). *Ind. Eng. Chem. Anal. Ed.*, **12,** 19–23.
2. Hirst, A. J. (1961). In: *The Applied Science of Rubber*, W. Naunton (ed.), London, Edward Arnold Ltd, pp. 587–708.
3. Lindley, P. B. and Stevenson, A. (1981). In: *Materials, Experimentation and Designs in Fatigue*, F. Sherratt and J. Sturgeon (eds), pp. 233–46.
4. Lake, G. J. and Lindley, P. B. (1964). *Rubber J.*, **146,** 24–30.
5. Lindley, P. B. and Teo, S. C. (1979). *Plast. Rubb: Mat. and Appl.*, **4,** 29–37.

6. Rivlin, R. S. and Thomas, A. G. (1953). *J. Polym. Sci.*, **10**, 291–318.
7. Lindley, P. B. (1972). *J. Strain Anal.*, **7**, 132–140.
8. Greensmith, H. W. (1963). *J. Appl. Polym. Sci.*, **7**, 993–1002.
9. Stevenson, A. (1983). *Int. J. Fracture*, In press.
10. Gent, A. N. (1956). *Proc. Rubber in Eng. Conf.*, London, MRDB, pp. 1–29.
11. Sedillot, F. and Stevenson, A. (1982). *Proc. 14th Offshore Technology Conference*, Houston.

Index